Lecture Notes in Mathematics 1580

Editors:
A. Dold, Heidelberg
B. Eckmann, Zürich
F. Takens, Groningen

Mario Milman

Extrapolation and
Optimal Decompositions

with Applications to Analysis

Springer-Verlag

Berlin Heidelberg New York
London Paris Tokyo
Hong Kong Barcelona
Budapest

Author

Mario Milman
Department of Mathematics
Florida Atlantic University
Boca Raton, FL 33431, USA
E-mail: Milman @acc.fau.edu

Mathematics Subject Classification (1991): Primary: 46M35, 46E35, 42B20
Secondary: 35R15, 58D25

ISBN 3-540-58081-6 Springer-Verlag Berlin Heidelberg New York
ISBN 0-387-58081-6 Springer-Verlag New York Berlin Heidelberg

CIP-Data applied for

© Springer-Verlag Berlin Heidelberg 1994
Printed in Germany

Typesetting: Camera-ready by author
SPIN: 10130043   46/3140-543210 - Printed on acid-free paper

To Vanda

Preface

In these notes we continue the development of a theory of extrapolation spaces initiated in [57]. One of our main concerns has been to connect the fundamental processes associated with the construction of interpolation and extrapolation spaces "optimal decompositions" with a number of problems in analysis. In particular we study extrapolation of inequalities related to Sobolev embedding theorems, higher order logarithmic Sobolev inequalities, *a.e.*convergence of Fourier series, bilinear extrapolation of estimates in different settings with applications to PDE's, apriori estimates for abstract parabolic equations, commutator inequalities with applications to compensated compactness, a functional calculus associated with positive operators on a Banach space, and the iteration method of Nash/Moser to solve nonlinear equations.

Many of the results presented in these notes are new and appear here for the first time. We have also included a number of open problems throughout the text. While we hope that these features could make our work attractive to specialists in the field of function spaces it is also hoped that the central rôle that the applications play in our development could also make it of interest to classical analysts working in other areas. In order to facilitate the task of these prospective readers we have tried to provide sufficient background information with complete references and included a brief guide to the literature on interpolation theory. We have also tried to make the contents of different chapters as independent of each other as possible while at the same time avoiding too much repetition. Finally we have also included a subject index and notation index.

It is a pleasure to record here my gratitude to a number of people and institutions who have helped me to complete these notes over

viii

the years. In particular I am grateful to Björn Jawerth with whom I have spend a lot of time over the years talking about extrapolätion. My friends DMGSJ played also an important extrapolatory role, both professionally and otherwise. The first version of the book was written during a membership, partially supported by an NSF grant, at the Institute for Advanced Study. I am most grateful to the Institute and Professor L. Caffarelli, for their support and for providing such an stimulating environment for my work. The notes were further developed while I was visiting the University of Paris (Orsay), The Centre for Ricerca (Barcelona) and the University of Zurich. I am particularly grateful to Professors Herbert Amann (Zurich), Aline Bonami (Orsay), Joan Cèrda (Barcelona) for their support and interest in my work.

Contents

1 Introduction   1
   1.1   **A Very Brief Guide To The Literature On Interpolation** 4

2 **Background On Extrapolation Theory**   7
   2.1   Introduction. 7
   2.2   More About the $\sum$ and Δ methods. 17
   2.3   Recovery of End Points 25
   2.4   The classical Setting of Extrapolation 28
   2.5   Weighted Norm Inequalities 31
   2.6   More Computations of Extrapolation Spaces 32
   2.7   Notes and Comments 32

3 **K/J Inequalities and Limiting Embedding Theorems**   35
   3.1   K/J Inequalities and Zafran Spaces 36
   3.2   Applications: Sobolev Imbeddings 39

4 **Calculations with the Δ method and applications**   43
   4.1   Reiteration and the Δ method 43
   4.2   On the Integrability of Orientation Preserving Maps . 46
      4.2.1   Background 47
      4.2.2   Identification of Sobolev Classes using Δ . . . 49
   4.3   Some Extreme Sobolev Imbedding Theorems 56

5 **Bilinear Extrapolation And A Limiting Case of a Theorem by Cwikel**   59
   5.1   Bilinear Extrapolation 60
   5.2   Ideals of Operators 67

    5.3   Limiting case of Cwikel's estimate   69
    5.4   Notes and Further Results   71

6 Extrapolation, Reiteration, and Applications   75
    6.1   Reiteration   75
    6.2   Estimates for the Maximal Operator Of Partial Sums
          Of Fourier Series   84
    6.3   Extrapolation Methods   85
    6.4   More On Reiteration   88
    6.5   Higher Order Logarithmic Sobolev Inequalities   89
    6.6   Notes and Further Results   91

7 Estimates For Commutators In Real Interpolation   95
    7.1   Some Operators Associated to Optimal Decompositions   97
    7.2   Method of Proof   102
    7.3   Computation of Ω for extrapolation spaces   107
    7.4   Other Operators Ω   111
    7.5   Compensated Compactness   115
    7.6   Relationship to Extrapolation   118
    7.7   A Functional Calculus   121
    7.8   A Comment on Calderón Commutators   124
    7.9   Notes and Further Results   125

8 Sobolev Imbedding Theorems and Extrapolation of
**   Infinitely Many Operators   127**
    8.1   Averages of Operators   127

9 Some Remarks on Extrapolation Spaces and Abstract
**   Parabolic Equations   131**
    9.1   Maximal Regularity   132

10 Optimal Decompositions, Scales, and Nash-Moser It-
**   eration   139**
    10.1 Moser's Approach to Solving NonLinear Equations .   140
    10.2 Scales with Smoothing and Interpolation   142
    10.3 Abstract Nash-Moser Theorem   145

Bibliography   149

Index   158

Symbols   161

Chapter 1

Introduction

In the last thirty years or so interpolation theory has become an important chapter in the field of function spaces. The origins of the theory are the classical interpolation or convexity theorems of Riesz, Thorin, and Marcinkiewicz. These classical results were subsequently extended by many authors including Calderón, Cotlar, Salem, Zygmund. The foundations of the general theory were laid down in the sixties by Aronszajn, Calderón, Gagliardo, Krein, Lions, Peetre, among others. It has since been extended, perfected, and applied, by many mathematicians. We refer the reader to the section at the end of this chapter for a brief guide to the available literature.

The theory has found many applications to classical analysis. In particular it has become an important tool in theories as diverse as partial differential equations, approximation theory, harmonic analysis, numerical analysis, operator theory, etc.

In the last few years, mainly in collaboration with B. Jawerth, we have been developing a new theory of "extrapolation spaces" which somehow is the converse of interpolation theory. The classical framework of interpolation theory can be briefly described as follows. We are given a pair of compatible Banach spaces (X, Y) and we attempt to construct all the spaces with the interpolation property between them. The real and complex methods, for example, provide parameterized families of spaces $(X, Y)_{\theta,q}$, $[X, Y]_\theta$, with the interpolation property. In extrapolation we conversely ask: given a family of interpolation spaces can we reconstruct the originating pair? This question is, of course, directly related to best possible interpolation theorems. In practice, however, one is also very much interested in weaker formulations of this general question. Thus, given a family of estimates for an operator T acting on interpolation spaces, one wishes to "extrapolate" this information either as an "extrapolation theorem" (i.e. a continuity result for T, the model for which is provided by the classical extrapolation theorem of Yano) or an "extrapolation inequality" for T where the extrapolation estimate is usually based on the basic functionals of real interpolation (K, J, E functionals). In a sense one could also consider "extrapolation" as chapter of the theory of interpolation of infinitely many spaces, since one is trying to obtain information from an infinite family of spaces. The precise connection between these theories is an interesting open problem.

The usual relationship between modern analysis and classical analysis is that the former provides a framework for consolidation, extension, and simplification of results of the latter. At times, however, it is the general framework that suggests the right questions, and the techniques of modern analysis sometimes also provide the answers. It is this symbiotic state of affairs that, from our perspective, makes the general field of interpolation theory interesting.

In these notes we study the connection between optimal decompositions, extrapolation, and its applications to other areas of analysis. In particular we also explore the role that cancellation plays in interpolation/extrapolation theory. We have arranged the development of the theory vis a vis concrete applications to classical analysis in order to emphasize our point of view. Thus the development of theory in this book is not "linear." A prospective reader interested mainly in the abstract theory could certainly skip the applications developed in each of the chapters. On the other hand we hope that the specific applications will be of interest to analysts working in different fields. As a consequence we have tried to make the reading of each of these sections as independent as possible from the others while at the same time trying to avoid too much repetition.

Most of the results presented in these notes are new or have not appeared in book form before. It is hoped that they will serve as the basis of a larger, more detailed, and formal book. The author would therefore welcome suggestions, remarks and corrections.

The notes are organized as follows. In Chapter 2 we provide a detailed introduction to extrapolation theory (the reader is referred to [57] and [58] for further information and other specific applications). In this chapter we also provide detailed computations of extrapolation spaces for different scales. In Chapter 3 we discuss K/J inequalities in the context of limiting embedding theorems for interpolation scales with specific applications to Sobolev embeddings, in particular we relate these results to limiting inequalities by Kato and Ponce [63] , and Beale, Kato and Majda [5] concerning singular integrals. In Chapter 4 we study the Δ method of extrapolation, develop new tools to compute it, and apply these results to sharpen recent estimates by Müller [82] and Iwaniec and Sbordone [54] on the integrability of the Jacobian of orientation preserving maps. We also give an application to Sobolev imbedding theorems extending recent work by Fusco, P. L. Lions and Sbordone [42]. In Chapter 5 we study bilinear extrapolation and prove general bilinear extrapolation theorems of Yano type. As an application, involving the Schatten classes, we derive an end point inequality, due to Constantin [22], of a well known theorem of Cwikel [28] concerning the singular values of Schrodinger operators (which leads to Constantin's [22] limiting version of the collective Sobolev estimates of Lieb [70]). In Chapter 6 we give a number of new reiteration theorems for extrapolation spaces and apply our results to give a new approach to end point estimates for the maximal operator of Fourier partial sums due to Sjölin [95] and Soria [97]. In this chapter we also provide a new approach to the higher order logarithmic Sobolev estimates of Gross and Feissner (cf. [46], [41]). In Chapter 7 we study the role of cancellation in interpolation/extrapolation theory: we prove new commutator theorems and establish the relationship between commutator theorems and a functional calculus for positive operators on a Banach space. We believe that this connection could lead to applications in the theory of parabolic equations. In this chapter we also develop in detail an application to the theory of compensated compactness of Murat and Tartar, and, in particular, we indicate how commutator theorems can be used to obtain sharp integrability theorems for Jacobians of orientation preserving maps. In the short Chapter 8 we consider families of operators an establish an abstract version of a Sobolev embedding theorem due to Varopoulos [102]. In

Chapter **9** we indicate rather briefly the role of extrapolation spaces in the theory of abstract parabolic equations in Banach space and establish a number of new end point estimates. Finally, in Chapter **10**, we discuss the iteration process of Nash/Moser in the setting of scales with smoothing and interpolation /extrapolation spaces, in particular we establish a precise relationship between the theory of scales of spaces with smoothing and interpolation scales as well as provide an interpretation of the paracommutators of Hörmander [50] in terms of optimal decompositions.

1.1   A Very Brief Guide To The Literature On Interpolation

This guide is intended to provide a rather incomplete and super brief guide to background literature on interpolation theory relevant to the developments in these notes. It is intended to help a newcomer to the field to begin to study the literature. We apologize in advance if your favorite book or paper is not quoted here. The classical reference books on interpolation theory include [13], [8], [101], and [67]. These books also develop many of the early applications of the theory to harmonic analysis, approximation theory, semigroups and partial differential equations. The book [65] also presents a detailed account of interpolation and its applications to operator ideals. More recent contributions are the research monograph [85] which develops a new approach to interpolation theory, the book [7] which presents a detailed study of rearrangement invariant spaces and the computation of K functionals, and the ground breaking treatise [12] presenting new important theoretical developments in the field.

The monograph [57] presents the first detailed account of extrapolation theory and contains a bibliography of 60 items, as well as brief historical survey. Further results and applications of extrapolation are contained in [58] (which also contains an extensive bibliography) and the forthcoming paper [27] which deals with complex extrapolation.

The theory of second order and commutator estimates in interpolation theory and their applications was developed in, among other papers, [89], [60], [25], [62], [61]. Finally we should mention that the recent book [26] that contains an extensive list of unsolved problems in the area.

Chapter 2

Background On
Extrapolation Theory

In this chapter we present in detail some of the basic results of the
theory developed in [57], [58]. In order to make the presentation
self contained we have included proofs to all the results presented.
For the benefit of the reader we have also provided a rather brief
introduction to interpolation theory.

2.1 Introduction.

We shall start by briefly recalling some basic definitions of interpolation theory.

We say that a pair of Banach spaces $\bar{A} = (A_0, A_1)$ is *compatible*
if there is a topological Hausdorff space $\mathcal{H}$ such that both A_0 and A_1
are subspaces of $\mathcal{H}$. If $\bar{A} = (A_0, A_1)$ is a pair of compatible spaces,
then a space A is called an *intermediate space* between A_0 and A_1 if

$$\Delta(\bar{A}) = A_0 \cap A_1 \subset A \subset \Sigma(\bar{A}) = A_0 + A_1$$

The spaces A and B are called *interpolation spaces* with respect to
$\bar{A}$ and $\bar{B}$ if for any bounded linear operator such that $T : \bar{A} \to \bar{B}$
(which simply means $T : A_i \to B_i$, $i = 0, 1$) we can conclude that
$T : A \to B$. Moreover, if $\|T\|_{A \to B} \leq \max\{\|T\|_{A_0 \to B_0}, \|T\|_{A_1 \to B_1}\}$
then we say that A and B are *exact*. An *interpolation method* $\mathcal{F}$
is a functor defined on compatible pairs of Banach spaces such that

if $\bar{A}$ and $\bar{B}$ are two pairs, then $\mathcal{F}(\bar{A})$ and $\mathcal{F}(\bar{B})$ are interpolation spaces with respect $\bar{A}$ and $\bar{B}$, and $\mathcal{F}(T)=T$ for all $T : \bar{A} \to \bar{B}$. A method $\mathcal{F}$ which yields exact interpolation spaces is called exact; if moreover, for some fixed $\theta \in (0,1)$ we have

$$\|T\|_{\mathcal{F}(\bar{A}) \to \mathcal{F}(\bar{B})} \le \left(\|T\|_{A_0 \to B_0} \right)^{1-\theta} \left(\|T\|_{A_1 \to B_1} \right)^{\theta}$$

for all $T : \bar{A} \to \bar{B}$, then we say that $\mathcal{F}$ is *exact of exponent* θ.

In extrapolation theory we start with families $\{A_\theta\}_{\theta \in \Theta}$ of Banach spaces indexed by some fixed index set Θ, (usually $\Theta = (0,1)$), which are *strongly compatible* in the sense that there are two Banach spaces Δ and Σ such that $\Delta \subset A_\theta \subset \Sigma$, $\theta \in \Theta$ (with continuous inclusions). Suppose now that $\{A_\theta\}_{\theta \in \Theta}$ and $\{B_\theta\}_{\theta \in \Theta}$ are families of strongly compatible spaces: $\Delta_a \subset A_\theta \subset \Sigma_a$ and $\Delta_b \subset B_\theta \subset \Sigma_b$ for some spaces Δ_a, Σ_a and Δ_b, Σ_b, respectively. In this setting the natural morphisms are the bounded, linear operators $T : \{A_\theta\}_{\theta \in \Theta} \xrightarrow{1} \{B_\theta\}_{\theta \in \Theta}$; this means that there is an operator $T : \Sigma_a \to \Sigma_b$, whose restriction to A_θ maps A_θ into B_θ, with norm ≤ 1 for each $\theta \in \Theta$. We say that the spaces A and B are *extrapolation spaces* (with respect to the families $\{A_\theta\}_{\theta \in \Theta}$ and $\{B_\theta\}_{\theta \in \Theta}$) if $\Delta_a \subset A \subset \Sigma_a$, $\Delta_b \subset B \subset \Sigma_b$, and

$$T : \{A_\theta\}_{\theta \in \Theta} \xrightarrow{1} \{B_\theta\}_{\theta \in \Theta} \implies T : A \to B$$

(or, more precisely, that T has an extension which is defined on A and maps A boundedly into B).

An extrapolation method $\mathcal{E}$ is a functor, defined on a collection $dom(\mathcal{E})$ of families of strongly compatible spaces, such that $\mathcal{E}(\{A_\theta\}_{\theta \in \Theta})$ and $\mathcal{E}(\{B_\theta\}_{\theta \in \Theta})$ are extrapolation spaces if $\{A_\theta\}_{\theta \in \Theta}$, and $\{B_\theta\}_{\theta \in \Theta} \in dom(\mathcal{E})$. In analogy with interpolation theory one can also define *exact extrapolation spaces* (rep. functors); in this case we require that

$$\|T\|_{A \to B} \le \sup_{\theta} \|T\|_{A_\theta \to B_\theta}.$$

The simplest extrapolation functors are the Σ and Δ methods. Let us now describe their natural domains. Suppose that $\{A_\theta\}_{\theta \in \Theta}$ is a family of strongly compatible spaces such that the norms $M_\Sigma(\theta)$ of the inclusions $A_\theta \to \Sigma$ are uniformly bounded,

$$\sup_{\theta} M_\Sigma(\theta) = \sup_{\theta} \frac{\|a\|_\Sigma}{\|a\|_{A_\theta}} < \infty . \tag{2.1}$$

Then we can form the *sum* $\Sigma_\theta(A_\theta)$ of the family $\{A_\theta\}_{\theta\in\Theta}$. $\Sigma_\theta(A_\theta)$ is the Banach space of all x in Σ which have a representation $x = \Sigma_\theta\, a_\theta$ (with absolute convergence in Σ), where $a_\theta \in A_\theta$, $\theta \in \Theta$, such that

$$\|x\|_{\Sigma(A_\theta)} = \inf\{\sum_\theta \|a_\theta\|_{A_\theta} : x = \Sigma_\theta a_\theta \} < \infty . \qquad (2.2)$$

The hypothesis (2.1) implies that $\Delta_\theta (A_\theta) \subset \Sigma_\theta (A_\theta) \subset \Sigma$.

It can be shown that if $\{A_\theta\}_{\theta\in\Theta}$ is an *ordered scale* (*i.e.* $\Theta = (0,1)$, and $A_{\theta_1} \overset{1}{\subset} A_{\theta_2}$ whenever $\theta_1 > \theta_2$), and $g(\theta)$ a positive decreasing function defined on $(0,1)$, then to compute the spaces $\Sigma_\theta(g(\theta)A_\theta)$ it is enough to consider the values of θ in a fixed sequence $\{\theta_\nu\} \subset \Theta$. To illustrate this, we prove it in detail for the case where $g(\theta) = \theta^{-\alpha}$ for some fixed $\alpha > 0$. Thus, we shall show that

$$\sum_\theta \left(\theta^{-\alpha}A_\theta\right) = \sum_{\nu=1}^\infty 2^{\nu\alpha} A_{2^{-\nu\alpha}}$$

Indeed, let $a \in \Sigma_\theta(\theta^{-\alpha}A_\theta)$, then there exists a representation of $a = \sum_{\nu=1}^\infty a_{\theta_\nu}$, with $a_{\theta_\nu} \in A_{\theta_\nu}$, and such that

$$\|a\|_{\sum_\theta(\theta^{-\alpha}A_\theta)} \approx \sum_{\nu=1}^\infty \theta_\nu^{-\alpha} \|a_{\theta_\nu}\|_{A_{\theta_\nu}}.$$

For each $\mu = 2, 3,...$, let $E_\mu = \{\nu : \theta_\nu \in [2^{-\mu}, 2^{-\mu+1})\}$, and $E = \{\nu : \theta_\nu \notin \bigcup_{\mu=2}^\infty E_\mu\}$. Let

$$b_{2^{-\mu}} = \begin{cases} \sum_{\nu\in E_\mu} a_{\theta_\nu}, & \mu = 2, ... \\[2mm] \sum_{\nu\in E} a_{\theta_\nu}, & \mu = 1 \end{cases}$$

then,

$$a = \sum_{\mu=1}^\infty b_{2^{-\mu}}$$

The norm of the first term is easily estimated using the triangle inequality and the ordering of the scale:

$$2^{-\alpha} \|b_{2^{-1}}\|_{A_{2^{-1}}} \le \sum_{\nu\in E} \theta_\nu^{-\alpha}\|a_{\theta_\nu}\|_{A_{2^{-1}}} \le \sum_{\nu\in E} \theta_\nu^{-\alpha}\|a_{\theta_\nu}\|_{A_{\theta_\nu}}.$$

Therefore,
$$\|b_{2^{-1}}\|_{\sum_\theta(\theta^{-\alpha}A_\theta)} \le c\|a\|_{\sum_\theta(\theta^{-\alpha}A_\theta)}$$
Similarly,

$$\sum_{\mu=2}^{\infty} 2^{\mu\alpha}\|b_{2^{-\mu}}\|_{A_{2^{-\mu}}} \le \sum_{\mu=2}^{\infty} 2^{\mu\alpha} \sum_{\nu\in E_\mu} \|a_{\theta_\nu}\|_{A_{\theta_\nu}}$$

$$= \sum_{\nu=1}^{\infty} \|a_{\theta_\nu}\|_{A_{\theta_\nu}} \sum_{\{-\log_2\theta_\nu\le\mu<1-\log_2\theta_\nu\}} 2^{\mu\alpha} \le 2^\alpha\|a\|_{\sum_\theta(\theta^{-\alpha}A_\theta)}$$

as desired.

In particular, if $\{A_\theta\}$ is an ordered scale, and if $\theta_0 \in (0,1)$, then, with norm equivalence, we have

$$\sum_{\theta\in(0,\theta_0)} \left(\theta^{-\alpha}A_\theta\right) = \sum_{\theta\in(0,1)} \left(\theta^{-\alpha}A_\theta\right) \tag{2.3}$$

The dual, Δ method, can be defined for strongly compatible families such that the norms $M_\Delta(\theta)$ of the inclusions $\Delta \to A_\theta$ are uniformly bounded,

$$\sup \ M_\Delta(\theta) = \sup_\theta \frac{\|a\|_{A_\theta}}{\|a\|_\Delta} < \infty \ , \tag{2.4}$$

Under this assumption we let

$$\Delta(A_\theta) = \{a \in \bigcap_\theta A_\theta : \|a\|_{\Delta(A_\theta)} = \sup_\theta \ \|a\|_{A_\theta} < \infty \ \}. \tag{2.5}$$

Observe that (2.4) implies that $\Delta \subset \Delta(A_\theta) \subset \Sigma$.

It is not hard to see that Σ and Δ are extrapolation methods which, in a certain sense, are extremal. Moreover, they are dual extrapolation functors (cf. Chapter 4, (4.3)). Variants of the Σ method can be defined for compatible families of Banach spaces $\{A_\theta\}_{\theta\in\Theta}$ such that the norms $M_\Sigma(\theta)$ of the inclusions $A_\theta \to \Sigma$ satisfy $\left(\sum_\theta(M_\Sigma(\theta))^{p'}\right)^{1/p'} < \infty$, where $1 \le p \le \infty$, $1/p' + 1/p = 1$. Then we define $\Sigma_p(A_\theta)$ by requiring

$$\|a\|_{\Sigma_p(A_\theta)} = \inf\left\{\left(\sum_\theta \|a_\theta\|_{A_\theta}^p\right)^{1/p} : a = \sum_\theta a_\theta, \ a_\theta \in A_\theta\right\} < \infty$$

In particular, $\Sigma_1 = \Sigma$.

Remark. In what follows we shall assume, unless explicitly specified, that we are dealing with strongly compatible families of spaces. Thus, a family of spaces shall mean a **strongly compatible** family of spaces, and in particular a **Banach pair** or simply a **pair** $\bar{A}$ shall mean a compatible Banach pair.

A function ρ on the half line is quasi-concave if ρ is increasing and $\frac{\rho(t)}{t}$ decreases. These functions play an important role in interpolation theory (cf. [8]).

Let $\mathcal{F}$ be an interpolation functor, then its characteristic function is defined by the relationship (cf. [12]) ,

$$\mathcal{F}(C, \frac{1}{t}C) = \frac{1}{\rho(t)}C \ , t > 0 \ . \tag{2.6}$$

If $\mathcal{F}$ is exact, then ρ is quasi-concave, and if $\mathcal{F}$ is exact of exponent θ, then $\rho(t) = Ct^\theta$. Conversely, if $\rho(t)$ is quasi-concave, then there are several exact interpolation methods with characteristic function $\rho(t)$. Among these, $\bar{A}_{\rho,\infty;K}$ and $\bar{A}_{\rho,1;J}$ are extremal. Recall that $\bar{A}_{\rho,\infty;K}$ is defined to be the collection of all $a \in \Sigma(\bar{A})$ such that

$$\|a\|_{\bar{A}_{\rho,\infty;K}} = \sup_{t>0} \frac{K(t, a; \bar{A})}{\rho(t)} < \infty \ . \tag{2.7}$$

where the Peetre K functional is defined by

$$K(t, a; \bar{A}) = \inf_{a=a_0+a_1} \{\|a_0\|_{A_0} + t\|a_1\|_{A_1} : a_i \in A_i, i = 0, 1\}$$

The space $\bar{A}_{\rho,1;J}$ consists of all $a \in \Sigma(\bar{A})$ such that

$$\|a\|_{\bar{A}_{\rho,1;J}} = \inf \int_0^\infty \frac{J(t, u(t); \bar{A})}{\rho(t)} \frac{dt}{t} < \infty, \tag{2.8}$$

where the infimum is taken over all representations $a = \int_0^\infty u(t)\frac{dt}{t}$ (with convergence in $\Sigma(\bar{A})$, $u(t) : (0, \infty) \to \Delta(\bar{A})$ strongly measurable), and where the J functional is defined by

$$J(t, a; \bar{A}) = \sup\{\|a\|_{A_0}, t\|a_1\|_{A_1}\}.$$

These functors are extremal in the sense that for any exact interpolation method $\mathcal{F}$ with characteristic function $\rho(t)$, then (cf. [85], [12]).

$$\bar{A}_{\rho,1;J} \xrightarrow{1} \mathcal{F}(\bar{A}) \xrightarrow{1} \bar{A}_{\rho,\infty;K} \ . \tag{2.9}$$

It is instructive to outline a proof of (2.9) here. To prove the first embedding we fix $u \in \Delta(\bar{A})$ and define $T(\lambda) = \lambda u$ for $\lambda \in C$. Then, $\|T\|_{C \to A_i} = \|u\|_{A_i}, i = 0, 1$, and, since $\mathcal{F}$ is exact,

$$|\lambda| \, \|u\|_{\mathcal{F}(\bar{A})} = \|T\lambda\|_{\mathcal{F}(\bar{A})}$$

$$\leq \max\{\|u\|_{A_0}, t\, \|u\|_{A_1}\} \, \|\lambda\|_{\mathcal{F}(C, \frac{1}{t}C)} = J(t, u; \bar{A}) \frac{1}{\rho(t)} \, |\lambda| \ .$$

Therefore,

$$\|u\|_{\mathcal{F}(\bar{A})} \leq \frac{1}{\rho(t)} J(t, u; \bar{A}), \ \forall t > 0. \tag{2.10}$$

If $a \in \Sigma(\bar{A})$ has a representation $a = \int_0^\infty u(t) \frac{dt}{t}$, then the triangle inequality combined with (2.10) gives

$$\|a\|_{\mathcal{F}(\bar{A})} \leq \int_0^\infty \frac{J(t, u(t); \bar{A})}{\rho(t)} \frac{dt}{t}$$

and taking the infimum over all representations $a = \int_0^\infty u(t) \frac{dt}{t}$, proves the first embedding. To prove the second, let us fix $a \in \Sigma(\bar{A})$ and $t > 0$, and define an operator T as a "linearized version" of $K(t, a; \bar{A})$ as follows: let $T(a) = K(t, a; \bar{A})$ and extend T, using the Hahn-Banach theorem, to a continuous linear functional on $\Sigma(\bar{A})$ with $|Tx| \leq K(t, x; \bar{A})$. Then, $\|T\|_{A_i \to C} \leq t^i, i = 0, 1$, and, using the fact that $\mathcal{F}$ is exact, we get that $\|Tx\|_{\mathcal{F}(C, \frac{1}{t}C)} \leq \|x\|_{\mathcal{F}(\bar{A})}$. In particular, for $x = a$ this implies

$$\frac{K(t, a; \bar{A})}{\rho(t)} \leq \|a\|_{\mathcal{F}(\bar{A})}$$

Thus, taking the supremum over $t > 0$ proves the second embedding of (2.9).

We consider the classical set up of real interpolation as developed, for example, in [8]. Let $\bar{A}$ be a pair of Banach spaces. For $0 < \theta < 1$

and $1 \leq q \leq \infty$ we let $\bar{A}_{\theta,q;K}$ be the space of all $a \in \Sigma(\bar{A})$ for which

$$\|a\|_{\theta,q;K} = c_{\theta,q} \left(\int_0^\infty (t^{-\theta} K(t,a;\bar{A}))^q \frac{dt}{t} \right)^{1/q} < \infty$$

where $c_{\theta,q} = ((1-\theta)\theta q)^{1/q}$. The constant $c_{\theta,q}$ has been chosen so that the characteristic function of the $(.,.)_{\theta,q;K}$ method is exactly t^θ. Similarly, we let $\bar{A}_{\theta,q;J}$ be the space of all $a \in \Sigma(\bar{A})$ for which

$$\|a\|_{\bar{A}_{\theta,q;J}} = c_{\theta,q}^* \inf \left(\int_0^\infty (t^{-\theta} J(t,u(t);\bar{A}))^q \frac{dt}{t} \right)^{1/q} < \infty,$$

where the infimum is taken over all representations $a = \int_0^\infty u(t)\frac{dt}{t}$ (convergence in $\Sigma(\bar{A})$) with $u(t)$ strongly measurable functions taking values in $A_0 \cap A_1$) and where $c_{\theta,q}^* = ((1-\theta)\theta q')^{-1/q'}$, $1/q + 1/q' = 1$. Again the constant has been chosen so that the characteristic function of the $(.,.)_{\theta,q;J}$ method is t^θ.

With these normalizations we have the imbeddings (cf.[57])

$$\bar{A}_{\theta,1;J} \xrightarrow{1} \bar{A}_{\theta,q;J}, \ \bar{A}_{\theta,q;K} \xrightarrow{1} \bar{A}_{\theta,\infty;K}, \bar{A}_{\theta,q;K} \xrightarrow{1} \bar{A}_{\theta,r;K}, \ q \leq r. \quad (2.11)$$

Given an interpolation functor $\mathcal{F}$ we let $\mathcal{F}(\bar{A})^\circ$ denote the closure of $\Delta(\bar{A})$ in $\mathcal{F}(\bar{A})$. For a pair $\bar{A}$, the *Gagliardo completion* of A_j, $j = 0, 1$, which we denote by $\tilde{A}_j$, is the set of elements $a \in \Sigma(\bar{A})$ which are $\Sigma(\bar{A})$ limits of bounded sequences in A_j or, equivalently, for which

$$\|a\|_{\tilde{A}_j} = \sup_{t>0} \frac{K(t,a;\bar{A})}{t^j} < \infty.$$

We say that a pair $\bar{A}$ is mutually closed if $A_j = \tilde{A}_j, j = 0, 1$. For an element $a \in \Sigma(\bar{A})$, the *Gagliardo diagram* $\Gamma(a)$ of a is defined by

$$\Gamma(a) = \{(x_0, x_1) : \exists a_j \in A_j \ s.t. \ \|a_j\|_{A_j} \leq x_j, j = 0, 1, a = a_0 + a_1\}$$

It is readily seen that $\Gamma(a)$ is a convex set of R^2. Its boundary may contain a semi-infinite vertical segment and/or a semi-infinite horizontal segment. The remainder of the graph will be the graph of a decreasing convex function $x_1 = \varphi(x_0)$. As it is well known (cf.

[8]) Gagliardo diagrams are closely linked with the computation of K functionals. Indeed, let

$$D(a) = \partial(a) \cap \{(x_0, x_1) \in R^2 : x_j > 0, \; j = 0, 1\} \; ,$$

then, for each $t > 0$, $K(t, a; \bar{A}) = K(t)$ is the x_0 intercept of the tangent to $D(a)$ with slope $-1/t$, the corresponding x_1 intercept is $\frac{K(t)}{t}$. Conversely, each point on the graph of φ intersects with the tangent of slope $-1/t$ for some values of t determined by the right and left derivatives of φ at x_0, and thus $x_0 + tx_1 = K(t)$

We are now ready to discuss the strong form of the "Fundamental Lemma" or **SFL** (cf. [29], [12]).

Lemma 1 *("The fundamental lemma"). Suppose that $\bar{A}$ is a pair of mutually closed spaces. Then $a \in \Sigma(\bar{A})^\circ$ if and only if there is a representation $a = \int_0^\infty u(s)\frac{ds}{s}$ with*

$$\int_0^\infty \min\left\{1, \frac{t}{s}\right\} J(s, u(s); \bar{A}))\frac{ds}{s} \leq \gamma \, K(t, a; \bar{A}), \; t > 0 \; , \qquad (2.12)$$

for some universal constant γ.

Proof.   For the reader's convenience we indicate here the main steps of the proof given in [24] and refer to this paper for complete details. A different proof is given in [12] where references to other proofs can be found. For the classical form of the fundamental lemma the reader is referred to [8] . The approach of [24] sketched here provides the best known value of the constant γ. Namely, $\forall a \in \Sigma(\bar{A})^\circ$, $\forall \varepsilon > 0$, there exists a decomposition $a = \int_0^\infty u(s)\frac{ds}{s}$, such that

$$\int_0^\infty \min\left\{1, \frac{t}{s}\right\} J(s, u(s); \bar{A}))\frac{ds}{s} \leq \gamma(1 + \varepsilon)^2 K(t, a; \bar{A})$$

with $\gamma \leq (3 + \sqrt{2})$.

Let us also point out that the method of proof of [24] is a modification of the one originally given by Cwikel [29]. Referring to the notation introduced above let us define

$$x_\infty = \sup\{x : (x, y) \in D(a)\}, \; x_{-\infty} = \inf\{x : (x, y) \in D(a)\}$$

$$y_\infty = \sup\{y : (x,y) \in D(a)\}, \ y_{-\infty} = \inf\{y : (x,y) \in D(a)\}$$

From the geometrical considerations above it follows that

$$x_\infty = \sup_t K(t,a;\bar{A}) = \|a\|_{A_0}, \ x_{-\infty} = \lim_{t\to 0} K(t,a;\bar{A})$$

$$y_{-\infty} = \lim_{t\to\infty} \frac{K(t,a;\bar{A})}{t}, \ y_\infty = \lim_{t\to\infty} \frac{K(t,a;\bar{A})}{t} = \|a\|_{A_1}$$

We construct a sequence of points lying in $D(a)$ following an idea of Gagliardo. Fix an arbitrary point (x_0, y_0) on $D(a)$, and let $r = 1+\sqrt{2}$ (which just happens to be the optimal value in our calculation). For $n > 0$ construct (x_n, y_n) inductively so that one of the following two alternatives holds: either $x_n = rx_{n-1}$, and $y_n \leq r^{-1}y_n$ or $x_n \geq rx_{n-1}$ and $y_n = r^{-1}y_{n-1}$. The process must stop if for some n we have either $rx_{n-1} \geq x_\infty$ or $r^{-1}y_{n-1} \leq y_{-\infty}$. In this case we stop the construction of our sequence at $n-1$, and we let the counter $\nu_\infty = n$, otherwise we continue the process indefinitely and we set $\nu_\infty = \infty$. For $n < 0$ we proceed "backwards" and inductively construct a sequence (x_n, y_n) such that either $x_n = r^{-1}x_{n+1}$ and $r^{-1}y_n \leq y_{n+1}$ or $x_n \leq r^{-1}x_{n+1}$ and $r^{-1}y_n = y_{n+1}$. The process must stop if for some $n < 0$ either $r^{-1}x_{n+1} \leq x_{-\infty}$ or $ry_{n+1} \geq y_\infty$ holds in which case we set $\nu_{-\infty} = n$ and we do not need to define x_n and y_n. Otherwise we set $\nu_{-\infty} = -\infty$. To the sequence $\{(x_n, y_n)\}_{\nu_{-\infty}-1<n<\nu_\infty+1}$ we associate a decomposition of a as follows. Given $\varepsilon > 0$, for each n, $\nu_{-\infty} < n < \nu_\infty$ we can then find, by definition, a decomposition

$$a = a_n + a'_n,$$

such that

$$x_n \leq \|a_n\|_{A_0} \leq (1+\varepsilon)x_n$$

and

$$y_n \leq \|a'_n\|_{A_1} \leq (1+\varepsilon)y_n.$$

For $\nu_{-\infty}+1 < n < \nu_\infty$ we define $u_n = a_n - a_{n-1}$, we also let $u_{\nu_\infty} = a - a_{\nu_\infty-1}$, if $\nu_\infty < \infty$, and $u_{\nu_{-\infty}+1} = a_{\nu_{-\infty}+1}$, if $\nu_{-\infty}+1 > -\infty$, and complete the definition by letting $u_n = 0$ in the remaining positions, if any. By considering separately the sum over the negative integers, and the one over the positive ones it can be readily seen that $\sum_{-\infty}^{\infty} u_n = a$, in $\sum(\bar{A})$. We also have, by the triangle inequality, and the definitions, that $\|u_n\|_{A_0} \leq (1+\varepsilon)(1+r^{-1})x_n$, and $\|u_n\|_{A_1} \leq (1+\varepsilon)(1+r^{-1})y_{n-1}$. Using these estimates, and some careful analysis, the reader should be able to show that $\forall t > 0$,

$$\sum_{-\infty}^{\infty} \min\{\|u_n\|_{A_0}, t\|u_n\|_{A_1}\} \leq (1+\varepsilon)(3 + 2\sqrt{2})K(t, a; \bar{A})$$

(otherwise we refer to [24] for the complete details). This is essentially a "discrete" version of what we wanted to prove. The "continuous" version we require is obtained as follows. Define for $n \in Z$,

$$S_n = \{\nu \in Z : (1+\varepsilon)^n < \frac{\|u_n\|_{A_0}}{\|u_n\|_{A_1}} \leq (1+\varepsilon)^{n+1}\}, \quad v_n = \sum_{\nu \in S_n} u_\nu,$$

and finally let

$$u(t) = \sum_{-\infty}^{\infty} (\log(1+\varepsilon))^{-1} v_n \chi_{((1+\varepsilon)^n, (1+\varepsilon)^{n+1}]}(t).$$

Then,

$$\int_0^\infty u(t)\frac{dt}{t} = \sum_{-\infty}^{\infty} v_n = \sum_{-\infty}^{\infty} u_\nu = a.$$

Moreover,

$$\int_0^\infty \min\{1, \frac{t}{s}\} J(s, u(s); \bar{A})\frac{ds}{s} = \sum_{-\infty}^{\infty} \int_{(1+\varepsilon)^n}^{(1+\varepsilon)^{n+1}} \min\{1, \frac{t}{s}\} J(s, u(s); \bar{A})\frac{ds}{s}$$

$$\leq \sum_{-\infty}^{\infty} \min\{1, \frac{t}{(1+\varepsilon)^{n+1}}\} \sum_{\nu \in S_n} \min\{\|u_\nu\|_{A_0}, (1+\varepsilon)^{n+1}\|u_\nu\|_{A_1}\}$$

$$\leq \sum_{-\infty}^{\infty} (1+\varepsilon) \sum_{\nu \in S_n} \min\{\|u_\nu\|_{A_0}, t\|u_\nu\|_{A_1}\}$$

$$\leq (1+\varepsilon) \sum_{-\infty}^{\infty} \min\{\|u_n\|_{A_0}, t\|u_n\|_{A_1}\} \leq (1+\varepsilon)^2 (3 + 2\sqrt{2})K(t, a; \bar{A})$$

as desired. $\square$

2.2 More About the Σ and Δ methods.

It is easy to compute the Σ and Δ functors for certain scales of real interpolation spaces. For example, in the next result from [57] (a *Fubini type theorem*) the identification is easy because the Σ functor *commutes* with $(.,.)_{\rho,1;J}$ functors, while the Δ functor *commutes* with the $(.,.)_{\rho,\infty;K}$ functors.

Theorem 2 *Let* $\bar{A}$ *be a pair of spaces and* $\{\rho_\theta\}_{\theta \in \Theta}$ *a family of quasi-concave functions.*

  i) If the function $\rho(t) = \sup_\theta \rho_\theta(t)$ *is finite at a point (and hence at all points), then*

$$\Sigma_\theta(\bar{A}_{\rho_\theta,1;J}) = \bar{A}_{\rho,1;J}$$

  ii) If the function $\rho^*(t) = \inf_\theta \rho_\theta(t)$ *is non-zero at a point, then*

$$\Delta_\theta(\bar{A}_{\rho_\theta,\infty;K}) = \bar{A}_{\rho^*,\infty;K} .$$

Proof.   We prove only (i), the proof of (ii) is similar. Since $\|a\|_{\bar{A}_{\rho,1;J}} \leq \|a\|_{\bar{A}_{\rho_\theta,1;J}}$, $\theta \in \Theta$, we clearly have

$$\|a\|_{\bar{A}_{\rho,1;J}} \leq \|a\|_{\Sigma_\theta(\bar{A}_{\rho_\theta,1;J})}$$

On the other hand, combining

$$\|a\|_{\Sigma_\theta(\bar{A}_{\rho_\theta,1;J})} \leq \frac{J(t,a;\bar{A})}{\rho(t)}$$

with the triangle inequality gives the converse inequality. $\square$

In order to be able to compute these functors over other families of spaces we need to work a little bit harder.

Given a function $M(\theta)$, $\theta \in (0,1)$, we associate $\tilde{M}(\theta)$, the largest logarithmically convex minorant of $M(\theta)$, and the concave function $\tau(t) = \inf_\theta M(\theta)t^\theta$, with $\lim_{t\to 0} \tau(t) = 0$, $\lim_{t\to\infty} \frac{\tau(t)}{t} = 0$. Thus, τ has a representation (cf. [8])

$$\tau(t) = \int_0^\infty \min\{1, \frac{t}{r}\} d\mu(r)$$

and $d\mu(r)(= -rd\tau'(r))$ is called the representing measure of τ.

Example 3 *If $M(\theta) \equiv 1$, then $\tau(t) = \min(1, t)$ and $d\mu(r) = \delta_1(r)$ (δ_1 denotes the Dirac measure at $r = 1$). If $M(\theta) \approx \theta^{-\alpha_1}$ as $\theta \to 0$ and $\approx (1-\theta)^{-\alpha_2}$ as $\theta \to 1$ for some $\alpha_1, \alpha_2 > 0$, then*

$$\tau(t) \approx \begin{cases} \log^{\alpha_1} t & \text{for large } t \\ \\ t\log^{\alpha_2} \frac{1}{t} & \text{for small } t \end{cases} \qquad (2.13)$$

and consequently

$$d\mu(r) \approx \begin{cases} (\log^{\alpha_1 - 1} r)\frac{dr}{r} & \text{as } r \to \infty \\ \\ \left(\log^{\alpha_2 - 1} \frac{1}{r}\right) dr & \text{as } r \to \infty \end{cases} \qquad (2.14)$$

If one of the $\alpha_i's$ is zero it should be replaced by a δ measure. For example if $\alpha_1 = 0$, then we may replace the term corresponding to the behavior at infinity by $\delta_1(r)$.

Let $1 \le q(\theta) \le \infty$, $\frac{1}{q'} = 1 - \frac{1}{q}$, and let

$$n(\theta) = \int_0^\infty t^{-\theta}\tau(t)\frac{dt}{t}$$

$$m(\theta) = m_q(\theta) = q^{1/q}(q')^{1/q'}(1-\theta)\theta \int_0^\infty t^{-\theta}\tau(t)\frac{dt}{t}$$

Then, if we assume that $M(\theta)$ is *tempered*, in the sense that $M(2\theta) \approx M(\theta)$ for θ close to 0, and $M(1 - 2(1-\theta)) \approx M(\theta)$ for θ close to 1, we have the following (cf. [57])

Proposition 4 *With constants of equivalence independent of θ*

$$\tilde{M}(\theta) \approx m(\theta) \approx n(\theta)$$

Proof.   Since $\tilde{M}(\theta) = \sup_{t>0} t^{-\theta}\tau(t)$, we use (2.26) with $r = \infty, q = 1$, applied to $K(t) = \tau(t)$, and we obtain

$$\tilde{M}(\theta) \le (1-\theta)\theta\, n(\theta)$$

On the other hand since $M(\theta)$ is tempered we have, for $t > 0$,

$$\tau(t^2) \le c\min\{1, t\}\tau(t).$$

We deduce that

$$(1 - \theta)\theta \, n(\theta) \leq c \int_0^\infty t^{-\theta} \min\{1, t^{\frac{1}{2}}\} \tau(t^{\frac{1}{2}}) \frac{dt}{t}$$

$$\leq c\tilde{M}(\theta) \int_0^\infty t^{-\theta} \min\{1, t^{\frac{1}{2}}\} \frac{dt}{t} = c\tilde{M}(\theta).$$

Finally, the equivalence $\tilde{M}(\theta) \approx m(\theta)$ is now trivial since $1 \leq q^{1/q} q'^{1/q'} \leq 2$. $\square$

We are now ready to show a powerful extension of Theorem 2.

Theorem 5 *Let $\bar{A}$ be a mutually closed pair of Banach spaces, and suppose that $M(\theta)$ is a positive, tempered function, and let $\{\mathcal{F}_\theta\}_{0<\theta<1}$ be a family of interpolation methods such that the characteristic function of $\mathcal{F}_\theta$ is equal to t^θ. Then, $\forall t > 0$,*

$$\|x\|_{\sum_\theta(t^\theta M(\theta)\bar{A}_{\theta,1;J})} \approx \|x\|_{\sum_\theta(t^\theta M(\theta)\mathcal{F}(\bar{A}))}$$

Thus, in particular

$$\sum_\theta (M(\theta)\bar{A}_{\theta,1;J}) = \sum_\theta (M(\theta)\mathcal{F}(\bar{A})) \qquad (2.15)$$

Proof.   In view of (2.9) it is sufficient to prove that

$$\|x\|_{\sum_\theta(t^\theta M(\theta)\bar{A}_{\theta,1;J})} \leq c\|x\|_{\sum_\theta(t^\theta M(\theta)\bar{A}_{\theta,\infty;K})}$$

The first step is to observe that if $d\mu$ is the measure representing $\tau(t) = \inf_\theta M(\theta)t^\theta$, and the pair $\bar{A}$ is mutually closed, then **SFL** implies that

$$\|a\|_{\sum_\theta(t^\theta M(\theta)\bar{A}_{\theta,1;J})} \approx \int_0^\infty K(\frac{t}{r}, a; \bar{A}) d\mu(r) \qquad (2.16)$$

For our purposes here it is enough to show that the left hand side of (2.16) is dominated by the right hand side. For each fixed t, the functor $t^\theta M(\theta)\bar{A}_{\theta,1;J}$ has characteristic function equal to $\rho_{\theta,t}(s) = (\frac{s}{t})^\theta M(\theta)^{-1}$. Thus, by Theorem 2 we have

$$\|a\|_{\sum_\theta(t^\theta M(\theta)\bar{A}_{\theta,1;J})} \leq c \inf_{a=\int_0^\infty u(s)\frac{ds}{s}} \int_0^\infty J(s, u(s); \bar{A}) \inf_\theta \left\{\left(\frac{t}{s}\right)^\theta M(\theta)\right\} \frac{ds}{s}$$

$$\leq c \inf_{a=\int_0^\infty u(s)\frac{ds}{s}} \int_0^\infty J(s, u(s); \bar{A}) \int_0^\infty \min\{1, \frac{t}{rs}\} d\mu(r) \frac{ds}{s}$$

$$\leq c \inf_{a=\int_0^\infty u(s)\frac{ds}{s}} \int_0^\infty \int_0^\infty J(s, u(s); \bar{A}) \min\{1, \frac{t}{rs}\} \frac{ds}{s} d\mu(r)$$

$$\leq c \int_0^\infty K(\frac{t}{r}, a; \bar{A}) d\mu(r)$$

where the last inequality follows from the SFL.

With this estimate at hand we may now continue with

$$\int_0^\infty K(\frac{t}{r}, a; \bar{A}) d\mu(r) = t^\theta \int_0^\infty t^{-\theta} K(\frac{t}{r}, a; \bar{A}) d\mu(r)$$

$$\leq t^\theta \int_0^\infty \sup_{t>0}\{t^{-\theta} K(\frac{t}{r}, a; \bar{A})\} d\mu(r) = ct^\theta n(\theta) \|a\|_{\bar{A}_{\theta,\infty;K}}$$

$$\leq ct^\theta M(\theta) \|a\|_{\bar{A}_{\theta,\infty;K}}$$

where in the last step we have used Proposition 4, and the fact that $\tilde{M}(\theta) \leq M(\theta)$. Thus, we have obtained the following estimate

$$\|a\|_{\sum(t^\theta M(\theta)\bar{A}_{\theta,1;J})} \leq ct^\theta M(\theta) \|a\|_{\bar{A}_{\theta,\infty;K}}$$

Consequently, by the definition of $\sum$,

$$\|a\|_{\sum(t^\theta M(\theta)\bar{A}_{\theta,1;J})} \leq c \|a\|_{\sum(t^\theta M(\theta)\bar{A}_{\theta,\infty;K})}$$

as we wished to show. $\square$

Remark. In particular, Theorem 5 applies to $\mathcal{F}_\theta = [.,.]_\theta$, $\mathcal{F}_\theta = (.,.)_{\theta,q;K}, 1 \leq q \leq \infty$.

Remark. Let $\rho(t)$ be a quasi-concave function, then inequalities of
the form

$$K(t, Ta; \bar{A}) \leq \|a\|_{A_0} \rho(\frac{\|a\|_{A_0}}{\|a\|_{A_1}})$$

or

$$K(t, Ta; \bar{A}) \leq c\rho(\frac{t}{s}) J(s, a, \bar{A})$$

are termed K/J inequalities. These inequalities are basic limiting estimates in the theory. When combined with the representation theory of quasi-concave functions and the fundamental lemma they allow us to reverse the interpolation process. Let us briefly indicate how they occur naturally in the theory (cf. also Theorem 7 below). Let $\bar{A}$ and $\bar{B}$ be Banach pairs, and let $\{\rho_\theta\}_{\theta\in\Theta}$, $\{\sigma_\theta\}_{\theta\in\Theta}$ be two families of quasi-concave functions. Set, $\tau(s,t) = \inf_\theta \frac{\sigma_\theta(t)}{\rho_\theta(s)}$, then the following two conditions are equivalent for an operator T (cf. [57])

$$T : \bar{A}_{\rho_\theta,1;J} \xrightarrow{1} \bar{B}_{\sigma_\theta,\infty;K}, \ \theta \in \Theta$$

$$K(t, Ta; \bar{B}) \leq \tau(s,t)J(s,a; \bar{A}), \ s,t > 0.$$

We shall now illustrate these results with some important examples. For $\alpha \in R$, let

$$\rho_\alpha(t) = \begin{cases} \left(1 + \log\frac{1}{t}\right)^{-\alpha} & 0 < t \leq 1 \\ t & 1 < t < \infty \end{cases} \tag{2.17}$$

Suppose that $\bar{A}$ is a pair of mutually closed spaces. Then, using Theorem 2, we see that for $\alpha \geq 0$, we have, with equivalence of norms,

$$\Sigma_{0<\theta<1}(\theta^{-\alpha}\bar{A}_{\theta,1;J}) = \bar{A}_{\rho_\alpha,1;J}. \tag{2.18}$$

On the other hand, by Theorem 5, it follows that for $1 \leq q \leq \infty$,

$$\Sigma_{0<\theta<1}(\theta^{-\alpha}\bar{A}_{\theta,q;K}) = \Sigma_{0<\theta<1}(\theta^{-\alpha}\bar{A}_{\theta,1;J}) = \bar{A}_{\rho_\alpha,1;J} \tag{2.19}$$

Still, the actual explicit characterization of the spaces that appear to the right in (2.18) and (2.19) necessitates some further work. In particular it is important to rewrite the norms of these spaces in terms of more readily computable functionals such as K functionals. Recall that a Banach pair $\bar{A}$ is said to be **ordered** if $A_1 \subset A_1$. For an ordered pair $\bar{A}$, $\alpha \geq 0$, and $0 < q \leq \infty$, we let $\bar{A}_{(\alpha);K}$ be the space of all $a \in A_0$, such that

$$\|a\|_{(\alpha);K} = \int_0^1 [(1 + \log\frac{1}{t})^{\alpha-1}K(t,a; \bar{A})]\frac{dt}{t} < \infty \tag{2.20}$$

if $\alpha > 0$, and for $\alpha = 0$ we let $\bar{A}_{(\alpha);K} = A_0^\circ$. In a similar vein, we define the corresponding J spaces for ordered pairs $\bar{A}$. The space $\bar{A}_{(\alpha);J}$, $\alpha > 0$, is the space of all $a \in A_0$ for which

$$\|a\|_{(\alpha),J} = \inf \int_0^1 [(1 + \log\frac{1}{t})^\alpha J(t, u(t); \bar{A})]\frac{dt}{t} < \infty \tag{2.21}$$

where the infimum is taken over all representations $a = \int_0^1 u(t)\frac{dt}{t}$ with $u(t) : (0,1) \rightarrow A_1$ strongly measurable. When $\alpha = 0$, we let $\bar{A}_{(\alpha);J} = A_0^\circ$. If, moreover, $\bar{A}$ is mutually closed, then the **SFL** implies that

$$\bar{A}_{(\alpha);K} = \bar{A}_{(\alpha);J},$$

and in this case we shall drop the subscripts K and J. Note that when $\bar{A}$ is ordered, we may replace the integrals $\int_0^\infty$ in the definition of $\bar{A}_{\rho_\alpha,1;J}$ by $\int_0^\beta$ for any finite $\beta > 0$. In fact, when $\beta \geq 1$, we have equality of norms.

Returning to (2.18) and (2.19), we see that

$$\Sigma_{0<\theta<1}(\theta^{-\alpha}\bar{A}_{\theta,q;K}) = \bar{A}_{(\alpha)} \qquad (2.22)$$

with

$$\|a\|_{\bar{A}_{(\alpha)}} \approx \begin{cases} \int_0^1 (1+\log\frac{1}{s})^{\alpha-1}K(s,a;\bar{A})\frac{ds}{s}, & \alpha > 0 \\ \\ K(1,a;\bar{A})^\circ, & \alpha = 0 \end{cases} \qquad (2.23)$$

There is a dual version of this as well. If we let $\rho_\alpha^*(t) = \frac{t}{\rho_\alpha(t)}$, then by a direct calculation $\rho_\alpha^*(t) \approx \inf_\theta(1-\theta)^{-\alpha}t^\theta$ and we have (cf. [57])

$$\Delta_{0<\theta<1}((1-\theta)^\alpha\bar{A}_{\theta,\infty;K}) = \bar{A}_{\rho_\alpha,\infty;K} \qquad (2.24)$$

for $\alpha > 0$.

In fact, if $\bar{A}$ is mutually closed, then we also have

$$\Delta_{0<\theta<1}((1-\theta)^\alpha\bar{A}_{\theta,1;J}) = \bar{A}_{\rho_\alpha,\infty;K} \qquad (2.25)$$

(for a much more general result concerning the Δ method see Theorem 21 in Chapter 4).

In particular, using the fact that $K(t,f,L^1,L^\infty) = t\ f^{**}(t)$, we shall obtain precise characterizations of certain extrapolation spaces associated with L^p spaces. Let $\alpha \geq 0$, and let Ω be a probability space. Then,

$$\Sigma_{1<p<\infty}((\frac{1}{p-1})^\alpha L^p(\Omega)) = \Sigma_{1<p<\infty}((\frac{1}{p-1})^\alpha L(p,1)(\Omega)) \qquad (2.26)$$

$$= L(LogL)^\alpha(\Omega),$$

and,

$$\Delta_{1<p<\infty}(p^\alpha L(p,\infty)) = Exp\, L^{1/\alpha}. \tag{2.27}$$

More generally, and for use later on, lets us show how one can extend these calculations to more general Lorentz spaces (cf. [72]). For a concave function $\varphi : (0,1) \to R_+$, let $\Lambda_\varphi(0,1)$ be the Lorentz space defined by the norm

$$\|f\|_{\Lambda_\varphi} = \int_0^1 f^*(s)d\varphi(s)$$

For a pair of Lorentz spaces $\Lambda_{\varphi_1}, \Lambda_{\varphi_2}$, the K functional is computed in [71]

$$K(t,f,\Lambda_{\varphi_1},\Lambda_{\varphi_2}) = \int_0^1 f^*(s)d\min\{\varphi_1(s),t\varphi_2(s)\} \tag{2.28}$$

Using (2.28), we get,

$$\|f\|_{(\Lambda_{\varphi_1},\Lambda_{\varphi_2})_{(\alpha)}} = \int_0^1 (\log \frac{1}{t})^{\alpha-1}\int_0^1 f^*(s)d\min\{\varphi_1(s),t\varphi_2(s)\}\frac{dt}{t}$$

For example, if $\varphi_1(s) = s$, and $\varphi_2(s) = 1$, $s \neq 0$, $\varphi_2(0) = 0$, then $\Lambda_{\varphi_1} = L^1$, and $\Lambda_{\varphi_2} = L^\infty$, and (2.28) gives $K(t,f,L^1,L^\infty) = tf^{**}(t)$.

Let us show a computation of extrapolation spaces associated with Lorentz spaces which is going to be useful later.

Theorem 6 *(cf. [72])*

$$\sum_\theta \{\theta^{-1}L(LogL)^{1+\theta}(T)\} = LLogL(LogLogL)(T)$$

Proof. Let us first give a detailed proof of the known fact that

$$L(LogL)^{1+\theta}(T) = [LLogL(T), L(LogL)^2(T)]_\theta \tag{2.29}$$

with norm equivalence independent of θ.

Note that the spaces $L(LogL)^\beta(T), \beta \in [0,2]$, have absolutely continuous norms. Then, for $\theta \in [0,1]$, we have *isometrically* (cf. [14], page 125)

$$[LLogL(T), L(LogL)^2(T)]_\theta = [LLogL(T)]^{1-\theta}[L(LogL)^2(T)]^\theta \tag{2.30}$$

To compute the spaces appearing on the right hand side of (2.30) we use another technique of Calderón [14]. First observe that, since we are dealing with spaces on a finite measure space, only *large values* of the Young's functions involved are important. In our case we shall take $x \geq e^e$.

Using Calderón's notation we write $L(LogL)^\beta = A_\beta^{-1}(L^1)$, where $A_\beta(x) = x(\log x)^\beta, \beta \in [0,2]$, for $x \geq e^e$. Although an explicit formula of A_β^{-1} is not readily available, we have that $\varphi_\beta(x) = x(\log x)^{-\beta}$ is equivalent to the inverse for large x. More precisely, an elementary computation shows that

$$\frac{x}{9} \leq \varphi_\beta(A_\beta(x)) \leq x, \text{ for } x \geq e^e, \beta \in [0,2]. \qquad (2.31)$$

Then, according to [14] page 166, we have, with norm equivalence independent of $\theta \in [0,1]$,

$$\phi_\theta^{-1}(L^1) = \left(A_1^{-1}(L^1)\right)^{1-\theta} \left(A_2^{-1}(L^1)\right)^\theta$$

$$= [LLogL]^{1-\theta} \left[L(LogL)^2\right]^\theta \qquad (2.32)$$

where

$$\phi_\theta^{-1}(x) = \left(A_1^{-1}\right)^{1-\theta}(x)\left(A_2^{-1}\right)^\theta(x) \qquad (2.33)$$

Therefore, combining (2.33), and (2.31), we obtain, for large values of x, and $\theta \in [0,1]$,

$$\phi_\theta^{-1}(x) \approx \left(x(\log x)^{-1}\right)^{1-\theta} \left(x(\log x)^{-2}\right)^\theta$$

$$\approx x(\log x)^{-(1+\theta)} \qquad (2.34)$$

Consequently, combining (2.34) with (2.32), and then with (2.31) once again, we obtain (7.15).

Thus, by (2.18)

$$\sum_\theta \{\theta^{-1}L(LogL)^{1+\theta}(T)\} = (LLogL, L(LogL)^2)_{(1);K} =$$

$$= \{f/ \int_0^1 K(t, f; LLogL, L(LogL)^2)\frac{dt}{t} < \infty\}$$

Since the spaces $LLogL$ and $L(LogL)^2$ are Lorentz spaces the K functional can be computed using (2.28) and we get

$$K(t, f; LLogL, L(LogL)^2) \approx$$

$$\approx \int_0^1 f^*(s)d \min\{\int_0^s (1 + \log \frac{1}{u})du, t \int_0^s \left[1 + \log \frac{1}{u}\right]^2 du\}$$

It follows that, for $t < e^{-1}$,

$$K(t, f; LLogL, L(LogL)^2) \approx$$

$$\approx \int_0^{e^{-\frac{1-3t}{t}}} f^*(s)(1 + \log \frac{1}{s})ds + t \int_{e^{-\frac{1-3t}{t}}}^{e^{-1}} f^*(s) \left[1 + \log \frac{1}{s}\right]^2 ds$$

Integrating with respect to $\frac{dt}{t}$ yields

$$\|f\|_{(LLogL, L(LogL)^2)_{(1);K}} \approx \int_0^{e^{-1}} f^*(s)(1 + \log \frac{1}{s})(\log(\log \frac{1}{s}))ds,$$

from which it follows that

$$\|f\|_{(LLogL, L(LogL)^2)_{(1);K}} \approx \|f\|_{LLogL(Log(LogL))}$$

as desired. $\square$

We shall give a general version of Theorem 6 in Theorem 48 below.

2.3 Recovery of End Points

An important aspect of the theory developed in [57] are the so called K/J inequalities which not only can be used to identify extrapolation spaces but allow the recovery of the end points, that is they allow us to reverse the interpolation process. This part of the so called "recovery of end points" theory developed in [57] and [58].

We shall now introduce the idea of *complete families* of extrapolation functors. Let us say that a Banach pair $\bar{A}$ is *regular* if $\Delta(\bar{A})$ is dense in A_0 and A_1. We say that a family of interpolation methods $\{\mathcal{F}_\theta\}_{\theta \in \Theta}$ is *complete* if whenever $\bar{A}$ and $\bar{B}$ are pairs of mutually closed spaces and $\bar{A}$ is regular, it holds that

$$T : \mathcal{F}_\theta(\bar{A}) \xrightarrow{1} \mathcal{F}_\theta(\bar{B}), \forall \theta \in \Theta \implies T : \bar{A} \to \bar{B}$$

It turns out that complete families can be characterized by their action on one dimensional spaces. This is the content of the next result.

Theorem 7 *(cf. [57]) A family of interpolation methods $\{\mathcal{F}_\theta\}_{\theta \in \Theta}$ is complete if and only there exists an absolute constant such that $\forall t, s > 0$*

$$\inf_\theta \frac{\rho_\theta(t)}{\rho_\theta(s)} \leq \min\{1, \frac{t}{s}\} \tag{2.35}$$

where ρ_θ denotes the characteristic function of the functor $\mathcal{F}_\theta$.

Proof.   We shall only consider in detail here the sufficiency of (2.35) and refer to [57] for a proof of the necessity of this condition. What we use is the fact that if $a \in \Delta(\bar{A})$, then

$$\|a\|_{\bar{A}_{\rho_\theta,1;J}} \leq \frac{\|a\|_{A_0}}{\rho_\theta \left(\frac{\|a\|_{A_0}}{\|a\|_{A_1}} \right)}$$

Since $\rho_\theta(t)$ is increasing and $\frac{\rho_\theta(t)}{t}$ is decreasing we see that

$$\sup_\theta \frac{\rho_\theta(t) \|a\|_{A_0}}{\rho_\theta \left(\frac{\|a\|_{A_0}}{\|a\|_{A_1}} \right)} \leq J(t, a; \bar{A})$$

Combining these estimates we get

$$\|a\|_{\Delta(\rho_\theta(t)\bar{A}_{\rho_\theta,\infty;K})} = \sup_\theta \rho_\theta(t) \|a\|_{\bar{A}_{\rho_\theta,\infty;K}} \leq \sup_\theta \rho_\theta(t) \|a\|_{\bar{A}_{\mathcal{F}(A)}}$$

$$\leq \sup_\theta \rho_\theta(t) \|a\|_{\bar{A}_{\rho_\theta,1;J}} \leq J(t, a; \bar{A}) \tag{2.36}$$

Suppose now that $T : \mathcal{F}_{\rho_\theta}(\bar{A}) \xrightarrow{1} \mathcal{F}_{\rho_\theta}(\bar{B})$, and the pair $\bar{B}$ is mutually closed, then the condition (2.35) implies that

$$\|Ta\|_{\Delta(\rho_\theta(t)\bar{B}_{\rho_\theta,\infty;K})} = \sup_s \sup_\theta K(s, Ta; \bar{B}) \frac{\rho_\theta(t)}{\rho_\theta(s)} \geq$$

$$\frac{1}{C} \sup_s \frac{K(s, Ta; \bar{B})}{\min\{1, \frac{s}{t}\}} = \frac{1}{C} J(t, Ta; \bar{B})$$

which together with

$$\|Ta\|_{\Delta(\rho_\theta(t)B_{\rho_\theta,\infty;K})} \leq \|a\|_{\Delta(\rho_\theta(t)\bar{A}_{\rho_\theta,\infty;K})}$$

and (2.36) gives

$$J(t, Ta; \bar{B}) \leq cJ(t, a; \bar{A}).$$

Thus, since $\Delta(\bar{A})$ is dense in A_0 and A_1, we obtain $T : \bar{A} \to \bar{B}$. □

Corollary 8 *Any $\{\mathcal{F}_\theta\}_{0<\theta<1}$ family of interpolation methods, with each $\mathcal{F}_\theta$ exact of exponent θ, is complete.*

Proof. Compute

$$\inf_{0<\theta<1} \frac{\rho_\theta(t)}{\rho_\theta(s)} = \inf_{0<\theta<1} \frac{C_\theta t^\theta}{C_\theta s^\theta} = \min\{1, \frac{t}{s}\}, \quad s, t > 0,$$

and apply the previous theorem to conclude. □

An interesting application of Theorem 7, given in [58], gives a proof of the following version of a theorem of Stafney, which was originally proved for the complex method of interpolation.

Example 9 *(Stafney, cf. [8], [58]) Suppose that $\bar{A}$ is a mutually closed regular pair and $[A_0, A_1]_{\theta_0} = A_0$, for some $\theta_0 \in (0, 1)$, then $A_0 = A_1$. In fact this result holds for much more general functors.*

The argument of the proof of Theorem 7 shows that if $\{\mathcal{F}_\theta\}_{\theta\in\Theta}$ is a complete family of interpolation functors, with characteristic functions ρ_θ, then, for any Banach pair $\bar{A}$ which is mutually closed, we have ,

$$\|a\|_{\Delta(\rho_\theta(t)\mathcal{F}_\theta(\bar{A}))} \approx J(t, a; \bar{A})$$

$\forall a \in \Delta(\bar{A}), \forall t > 0.$

Dually, a similar argument yields that, for a complete family of interpolation functors $\{\mathcal{F}_\theta\}_{\theta\in\Theta}$, and every Banach pair $\bar{A}$, which is mutually closed and regular, we have

$$\|a\|_{\sum(\rho_\theta(t)\mathcal{F}_\theta(\bar{A}))} \approx K(t, a; \bar{A}) \tag{2.37}$$

$\forall a \in \sum(\bar{A}), \forall t > 0$ (cf. the proof of Theorem 5).

2.4   The classical Setting of Extrapolation

It is instructive to return to the classical setting of the extrapolation theorem of Yano, as developed in Zygmund's book [104].

Suppose that T is a bounded linear operator on $L^p(0,1)$ for $p > 1$ with $\|T\|_{L^p \to L^p} = \mathcal{O}((p-1)^{-\alpha})$, as $p \to 1$, for some $\alpha > 0$; then these estimates can be extrapolated to $T : L(LogL)^\alpha \to L^1$. There is also a dual statement for operators T acting on $L^p(0,1)$ for p close to ∞, with $\|T\|_{L^p \to L^p} = \mathcal{O}(p^\alpha)$, as $p \to \infty$, for some $\alpha > 0$; then $T : L^\infty \to Exp\, L^{1/\alpha}$.

To prove the first half of Yano's theorem, we apply the $\sum$ functor to obtain

$$T : \sum_{p>1}((p-1)^{-\alpha}L^p) \to L^1$$

and conclude by (2.26) that

$$T : L(LogL)^\alpha \to L^1$$

Similarly, the second half follows using the Δ functor and (2.27).

The reader will observe that, even in this classical setting, we have obtained a much more general result through the use of the abstract methods. For example, the assumptions of Yano's theorem imply that for any $\beta > 0$, we have

$$T : (p-1)^{-(\alpha+\beta)}L^P \to (p-1)^{-\beta}L^p$$

so that by extrapolation

$$T : \sum_{p>1}(p-1)^{-(\alpha+\beta)}L^P \to \sum_{p>1}(p-1)^{-\beta}L^p,$$

and consequently

$$T : L(LogL)^{\alpha+\beta} \to L(LogL)^\beta$$

This is part of a general theory of **"division of inequalities"** developed in [57].

We also observe that the classical proof of Yano's theorem, as given for example in Zygmund's book [104], does not use the full force of the hypothesis on the operator T. We shall now show that, if we insist on using all the information provided by the assumptions, we can recover, via extrapolation, very sharp "rearrangement inequalities" for classical operators.

Let $\bar{A}$ be a pair of mutually closed spaces, and let $d\mu$ be the representing measure of $\tau(t) = \inf_\theta M(\theta)t^\theta$, then as a consequence of **SFL** we have, as we have pointed before,

$$\|a\|_{\sum_\theta(t^\theta M(\theta)\bar{A}_{\theta,q(\theta),;K})} \approx \int_0^\infty K(\frac{t}{r}, a; \bar{A})d\mu(r).$$

For example, if $M(\theta) \approx \theta^{-1}$, as $\theta \to 0$, and $M(\theta) \approx (1-\theta)^{-1}$, as $\theta \to 1$, then

$$\|a\|_{\sum_\theta(t^\theta M(\theta)\bar{A}_{\theta,q(\theta),;K})} \approx \int_0^\infty K(s, a; \bar{A})\min\{1, \frac{t}{s}\}\frac{ds}{s} \qquad (2.38)$$

In the special case of rearrangement invariant spaces these methods lead to rearrangement inequalities for classical operators including the Hilbert transform or more generally Calderón-Zygmund operators, etc. For example, as is well known, the Hilbert transform is an example of an operator acting on $L^p(R^n)$ satisfying

$$\|Hf\|_p \le c\,\frac{p^2}{p-1}\|f\|_p, 1 < p < \infty, \qquad (2.39)$$

Since the pair (L^1, L^∞) is mutually closed, and the L^p spaces can be obtained by real interpolation

$$L^p = (L^1, L^\infty)_{1-\frac{1}{p}, p; K}$$

We see that, with $\theta = 1 - \frac{1}{p}$, we have

$$\|Hf\|_{(L^1, L^\infty)_{\theta, p(\theta); K}} \le \frac{c}{\theta(1-\theta)}\|f\|_{(L^1, L^\infty)_{\theta, p(\theta); K}}, \theta \in (0, 1)$$

By extrapolation, we get

$$\|Ha\|_{\sum_\theta(t^\theta(L^1, L^\infty)_{\theta, p(\theta); K})} \le \|a\|_{\sum_\theta(t^\theta M(\theta)(L^1, L^\infty)_{\theta, p(\theta); K})}$$

Using (2.37) we see that the left hand side is equivalent to $K(t, Ha; L^1, L^\infty)$, while the right hand side can be computed using (2.38). Thus, we have

$$K(t, Ha; L^1, L^\infty) \le c \int_0^\infty K(s, a; L^1, L^\infty) \min\{1, \frac{t}{s}\} \frac{ds}{s}$$

Now, using $K(t, Ha; L^1, L^\infty) = t(Ha)^{**}(t)$, $K(s, a; L^1, L^\infty) = sa^{**}(s)$, we finally obtain

$$t(Ha)^{**}(t) \le c \int_0^\infty s\, a^{**}(s)\, \min\{1, \frac{t}{s}\} \frac{ds}{s}$$

$$= c \left(\int_0^t a^{**}(s)ds + t \int_t^\infty a^{**}(s) \frac{ds}{s} \right) \tag{2.40}$$

The inequality (2.40) is a well known rearrangement inequality of Calderón-Stein-Weiss (cf. [57]). It is actually well known that it is possible to improve upon (2.40) using the fact that the Hilbert transform is of weak type (1,1) (cf. [57] for a much more general result).

More generally, if $M(\theta) \approx \theta^{-\alpha}(1 - \theta)^{-\beta}$, say, then using (2.14) we obtain

$$\|a\|_{\sum_\theta (t^\theta M(\theta) \bar{A}_{\theta, q(\theta), ; K})}$$

$$\approx \int_0^\infty K(s, a; \bar{A})\{(\log^+ \frac{t}{s})^{\alpha-1}) + (\log^+ \frac{s}{t})^{\beta-1}\} \min\{1, \frac{t}{s}\} \frac{ds}{s} \tag{2.41}$$

Thus, if T is an operator acting on L^p spaces, that satisfies

$$\|Tf\|_p \le c \frac{p^{\alpha+\beta}}{(p-1)^\alpha} \|f\|_p, \; 1 < p < \infty \tag{2.42}$$

We obtain the rearrangement inequality.

$$tT(f)^{**}(t) \le c \int_0^\infty a^{**}(s)\{(\log^+ \frac{t}{s})^{\alpha-1} + (\log^+ \frac{s}{t})^{\beta-1}\} \min\{1, \frac{t}{s}\} ds$$

The point is that the argument just presented works for operators acting in general scales of interpolation spaces. In this case the estimates are given in terms of K functionals.

2.5 Weighted Norm Inequalities

We develop further the idea that starting from a family of estimates, like (2.39), we can produce very precise estimates for a given operator, including weighted norm estimates.

For example, Sawyer [91] has given necessary and sufficient conditions on weights, for weighted L^p norm inequalities to hold for certain positive operators acting on decreasing functions. In particular, he has characterized the weights (w_0, w_1) such for a given $p \in (1, \infty)$, the operator

$$Sf(t) = \int_0^\infty f(s) \min\{\frac{1}{t}, \frac{1}{s}\} ds$$

satisfies

$$\|Sf\|_{L^p(w_1)} \leq c \|f\|_{L^p(w_0)}$$

for all f decreasing. Thus, if T is an operator satisfying norm estimates of the type (2.39), we get

$$\frac{K(t, Tf; L^1, L^\infty)}{t} \leq c \int_0^\infty \frac{K(s, f; L^1, L^\infty)}{s} \min\{\frac{1}{t}, \frac{1}{s}\} ds$$

$$= cS(\frac{K(s, f; L^1, L^\infty)}{s})(t)$$

Now, since $K(s)/s$ is decreasing, we see that, if the weights (w_0, w_1) satisfy Sawyer's conditions, we get

$$\int_0^\infty [\frac{K(t, Tf; L^1, L^\infty)}{t}]^p w_1(t) dt \leq c \int_0^\infty [\frac{K(s, f; L^1, L^\infty)}{s}]^p w_0(s) \frac{ds}{s}$$

Sawyer's results also give sharp conditions for pairs of weights (w_0, w_1) that imply weighted norm inequalities of $(L^p(w_0), L^p(w_1))$ type, for operators defined like the right hand side of (2.41) on decreasing functions.

Thus, the analysis presented here will also produce results in the situation described in (2.42). The same argument also applies to other weighted norm estimates for the operator S (e.g. $(L^p(w_0), L^q(w_1))$ estimates, etc.).

Finally we note that this argument is very general, and independent of the L^p scale of spaces, and produces very sharp extrapolation theorems.

2.6 More Computations of Extrapolation Spaces

We indicate here a few more computations of extrapolation spaces in the classical setting of L^p spaces, using *modern technology*.

In [43] we show that for an ordered pair $\bar{A}$ we have

$$K(t, f, A_0, \bar{A}_{(0);K}) \simeq t \int_{e^{-\frac{1}{t}}}^1 K(s, f, A_0, \bar{A}_1) \frac{ds}{s} \qquad (2.43)$$

It follows that

$$(A_0, \bar{A}_{(0);K})_{(0);K} = \{f : \int_0^1 K(s, f) \frac{1}{\ln \frac{1}{s}} \frac{ds}{s} < \infty\} \qquad (2.44)$$

Example 10 *Consider the pair $\bar{A} = (L^1, L^\infty)$ over a finite measure space. We have shown that $\bar{A}_{(0);K} = LLogL$. Iteration using (2.44) leads to*

$$(L^1, LLogL)_{(0);K} = \{f / \int_0^1 s f^{**}(s) \frac{1}{\ln \frac{1}{s}} \frac{ds}{s} < \infty\}$$

$$= LLog(LogL)$$

We can continue computing...

$$(L^1, LLog(LogL))_{(0);K} = LLog(Log(LogL))$$

2.7 Notes and Comments

For the most part the results in this chapter come from [57]. The counterpart of Theorem 5 for the Δ method will be established in Theorem 21. The connection with Sawyer's weighted norm inequalities seems to be new and complements earlier work in [90] on the use of weighted norm inequalities in interpolation theory. In the classical context of L^p spaces, different extrapolation methods have been used by Kerman [64] to obtain estimates like (2.40). For a detailed study of Lorentz-Zygmund spaces we refer to [6] and the references quoted therein.

Other aspects of the theory of K/J inequalities are developed in Chapter 3 in the context of embedding theorems, and in Chapter 5 in the context of bilinear extrapolation.

We have not said anything about non-linear extrapolation. It is easy to see that one can extrapolate operators that are $K-$ linear or quasi-linear in suitable technical senses. We now indicate less standard conditions under which we can extrapolate non-linear operators. In the context of lattices, Cwikel and Nilsson [30] have obtained versions of **SFL** where the functions in the representation have disjoint supports. A typical application to extrapolation is the following (cf. [27])

Theorem 11 *Let $\bar{A}$ be an ordered pair of mutually closed Banach lattices on a measure space $(\Omega, \sum, \mu)$. Let $\{\mathcal{F}_\theta\}_{\theta \in \Theta}$ be a family of exact interpolation functors, with each $\mathcal{F}_\theta$ of order θ. Let $\alpha > 0$, then there exists an absolute constant c, such that if $f \in \sum_\theta \theta^{-\alpha} \mathcal{F}_\theta(\bar{A})$, then, there exists a representation of f as a series, $f = \sum_{n=1}^{\infty} f_n$, where the functions f_n have disjoint supports, and*

$$\sum_{n=1}^{\infty} \theta_n^{-\alpha} \|f\|_{\mathcal{F}_{\theta_n}(\bar{A})} \leq c \|f\|_{\sum_\theta \theta^{-\alpha} \mathcal{F}_\theta(\bar{A})},$$

for some sequence of numbers $\{\theta_n\}$ in $(0,1)$.

This result can be used to prove extrapolation theorems for non-linear operators that are additive for disjointly supported combinations of functions. For example, we can readily prove the following

Corollary 12 *Let $\bar{A}$ be an ordered mutually closed pair of lattices, X be a Banach space, and $\{\mathcal{F}_\theta\}_{\theta \in \Theta}$ be a family of exact interpolation functors, with each $\mathcal{F}_\theta$ exact of order θ. Suppose that T is an (not necessarily linear) operator, $T : \mathcal{F}_\theta(\bar{A}) \to X$, $\forall \theta \in \Theta$, and such that*
(i) For some $\alpha > 0$, $\forall r > 0$, $\forall \theta \in \Theta$,

$$sup_{\|f\|_{\mathcal{F}_\theta(A)} \leq r} \|Tf\|_X \leq c\theta^{-\alpha} r.$$

(ii) if $f = \sum_{n=1}^{\infty} f_n$, and the $f_n's$ are disjointly supported, then $Tf = \sum_{n=1}^{\infty} Tf_n$

Then,

$$T : \sum_{\theta} \theta^{-\alpha} \mathcal{F}_{\theta}(\bar{A}) \to X$$

and there exists an absolute constant $c > 0$ such that

$$\sup_{\|f\|_{\sum_{\theta}(\theta^{-\alpha}\mathcal{F}_{\theta}(\bar{A}))} \leq r} \|Tf\|_X \leq cr$$

Example 13 *The following operators satisfy the conditions of the previous Corollary. Superposition operators $u \to f(x, u(x))$, such that $f(x, 0) = 0$, integral operators of the form*

$$Kf(t) = \int k(t, s, f(s)) ds, \text{ with } k(t, s, 0) = 0$$

NOTATION: We shall at times, when it is not important to be precise about constants, and the spaces are equivalent, drop the subscript K or J when dealing with real interpolation spaces. Whenever dealing with families of spaces we shall assume that they are strongly compatible unless otherwise specified.

Chapter 3

K/J Inequalities and Limiting Embedding Theorems

In this chapter we study K/J inequalities in connection with limiting embedding theorems. We focus on the replacement of the classical "power type interpolation inequalities" by analogues using quasiconcave functions, and the relationship of these inequalities with extrapolation.

In the classical theory of weak type estimates or more generally in classical real interpolation theory it is easy to characterize the continuity of an operator from spaces of the form $\bar{A}_{\theta,1;J}$ to any Banach space Z. We exploit the fact that the extrapolation spaces, obtained by applying the $\sum$ method to a family of real interpolation spaces, are of the form $\bar{A}_{\rho,1;J}$, in order to give a simple criteria for the continuity of mappings from extrapolation spaces to any Banach space. When we apply this criteria in the setting of embedding theorems we obtain estimates which are proving useful in the theory of partial differential equations. For example, we give a new approach to the results of Brezis-Gallouet [10], and Brezis-Wainger [11]. Also recent results concerning end point inequalities for singular integrals by Kato and Ponce [63], Beale, Kato, and Majda [5] can be treated in this fashion. More importantly, the methods developed here are very general and can be applied to other operators and other scales (e.g. Besov spaces associated to semigroups of operators, which we intend to discuss elsewhere.) Our presentation here has been influenced and complements results in the recent monograph by Taylor [99] where, among many other things, the uses of K/J inequalities and their applications to the theory of hyperbolic equations is presented.

3.1 K/J Inequalities and Zafran Spaces

Let $\bar{X}$ be a Banach pair, and let $\{X_\theta\}_\theta$ be a family of spaces. Let us consider the familiar "interpolation inequalities" of the form

$$\|x\|_\theta \leq c_\theta \|x\|_0^{1-\theta} \|x\|_1^\theta \tag{3.1}$$

It is a well known fact (cf. [8]) that (3.1) is actually equivalent to the embeddings

$$\bar{X}_{\theta,1;J} \subset X_\theta$$

Even in this classical setting there are interesting variants of (3.1) containing logarithmic factors. As a motivation let us look at a scale $\bar{A}_{\theta,q;K}$ of spaces obtained by real interpolation. Then, we can prove the following

$$\|x\|_{\bar{A}_{\theta,1;K}} \leq c_\theta \|x\|_{\bar{A}_{\theta,p;K}} (1 + \log^{1/p'} \frac{\|x\|_{A_1}}{\|x\|_{\bar{A}_{\theta,p;K}}}) \tag{3.2}$$

Indeed, for $\lambda \in (0,1)$, write

$$\|x\|_{\bar{A}_{\theta,1;K}} = c_\theta \left[\int_0^\lambda s^{-\theta} K(s,x;\bar{A}) \frac{ds}{s} + \int_\lambda^1 s^{-\theta} K(s,x;\bar{A}) \frac{ds}{s} \right]$$

Using the fact that $\sup_s \frac{K(s,x;\bar{A})}{s} \leq \|x\|_{A_1}$ to estimate the first term, and Hölder's inequality, and the definition of the $\bar{A}_{\theta,p;K}$ norm, to estimate the second term, we obtain

$$\|x\|_{\bar{A}_{\theta,1;K}} \leq c_\theta [\|x\|_1 \lambda^{1-\theta} + \|x\|_{\bar{A}_{\theta,p;K}} (\log \frac{1}{\lambda})^{1/p'}],$$

and choosing $\lambda = [\frac{\|x\|_{\bar{A}_{\theta,p;K}}}{\|x\|_{A_1}}]^{\frac{1}{1-\theta}}$, we obtain

$$\|x\|_{\bar{A}_{\theta,1;K}} \leq c_\theta \|x\|_{\bar{A}_{\theta,p;K}} (1 + \log^{1/p'} (\frac{\|x\|_1}{\|x\|_{\bar{A}_{\theta,p;K}}}))$$

Example 14 *Let Ω be a probability space, $A_0 = L^1(\Omega)$, $A_1 = L^\infty(\Omega)$, then $L(p,1)(\Omega) = (A_0, A_1)_{\frac{1}{p'},1;K}$, and we have*

$$\|f\|_{L(p,1)} \leq c\|f\|_{L^\infty}(1 + \log^{1/p'}(\frac{\|f\|_{L^1}}{\|f\|_{L^\infty}}))$$

Let $\bar{A}$ be an ordered pair, and let Z be a compatible Banach space, we consider K/J inequalities of the form

$$\|x\|_Z \leq c\|x\|_{A_0} f(\frac{\|x\|_{A_1}}{\|x\|_{A_0}})$$

in connection with extrapolation.

Associated with this type of estimate is a construction due to Zafran (cf. [103], [57]) which we now recall. Let $\bar{A}_f$ be defined as the closure of A_1 in A_0 with the norm $\|.\|_f$ given by

$$\|x\|_f = \inf\{\sum_{k=1}^n |a_k| f(\|x_k\|_{A_1}) : x = \sum_{k=1}^n a_k x_k, \text{ and } \sup_{1 \leq k \leq n} \|x_k\|_{A_0} \leq 1\} \tag{3.3}$$

Then, if $\rho(t) = \frac{1}{f(\frac{1}{t})}$, it is shown in [57] that

$$\bar{A}_{\rho,1;J} = \bar{A}_f \tag{3.4}$$

The only serious issue in the proof is the fact that if $x \in A_1$, then in the computation of the norm $\|x\|_{\bar{A}_{\rho,1;J}}$ we only need to consider decompositions with only a finite number of terms. More precisely, we have that if $x \in A_1$, then

$$\|x\|_{\bar{A}_{\rho,1;J}} \approx \inf\{\sum_{k=1}^n \frac{J(2^{-k}, x_k; \bar{A})}{\rho(2^{-k})} : x = \sum_{k=1}^n x_k, x_k \in A_1\}$$

We now record the following consequence of this result.

Theorem 15 *Let $\bar{A}$ be an ordered pair, and let $f : [0,\infty) \to [1,\infty)$ be a quasi-concave function such that $\lim_{t\to\infty} f(t) = \infty$, $\lim_{t\to\infty} \frac{f(t)}{t} = 0$. Let $\rho(t) = \frac{1}{f(\frac{1}{t})}$, and let X be a Banach space. Suppose that T is a bounded linear operator $T : A_1 \to X$, then T can be extended to a bounded operator $T : \bar{A}_{\rho,1;J} \to X$ if and only if there exists an absolute constant $c > 0$ such that $\forall x \in A_1$*

$$\|Tx\|_X \leq c\|x\|_{A_0} f(\frac{\|x\|_{A_1}}{\|x\|_{A_0}}) \tag{3.5}$$

Proof.   Given the equivalence (3.4) we have that T can be extended to $\bar{A}_{\rho,1;J}$ if and only if there exists a constant $c > 0$ such that $\forall x \in A_1$ we have $\|Tx\|_X \leq c\|x\|_f$. Now, suppose that (3.5) holds. Let

$$x = \sum_{k=1}^{n} a_k x_k, \text{ with } x_k \in A_1, \; \sup_{1 \leq k \leq n} \|x_k\|_{A_0} \leq 1$$

and

$$\sum_{k=1}^{n} |a_k| f(\|x_k\|_{A_0}) \approx \|x\|_f$$

Then,

$$\|Tx\|_X \leq \sum_{k=1}^{n} |a_k| \|Tx_k\|_X \leq \sum_{k=1}^{n} |a_k| \|Tx_k\|_X$$

$$\leq c \sum_{k=1}^{n} |a_k| \|x_k\|_{A_0} f(\frac{\|x_k\|_{A_1}}{\|x_k\|_{A_0}})$$

$$\leq c \sum_{k=1}^{n} |a_k| f(\|x_k\|_{A_1})$$

$$\approx \|x\|_f$$

where the last inequality follows using the quasi-concavity of f.

Conversely, suppose that $T : \bar{A}_{\rho,1;J} \to X$, then, in particular, there exists a constant $c > 0$ such that

$$\|Tx\|_X \leq c\|x\|_f, \forall x \in A_1$$

Now, the decomposition $x = \frac{\|x\|_{A_0} x}{\|x\|_{A_0}}$ shows that

$$\|x\|_f \leq \|x\|_{A_0} f(\frac{\|x\|_{A_1}}{\|x\|_{A_0}})$$

and therefore (3.5) holds. $\square$

Combining the previous theorem with the characterization of the spaces $\bar{A}_{\rho,1;J}$ as extrapolation spaces gives (cf. Chapter **2**, Theorem **2**).

Theorem 16 *Let (X,Y) be a mutually closed ordered pair, Z be a Banach space, and let $\{\mathcal{F}_\theta\}_{\theta \in (0,1)}$ be a family of interpolation functors such that each $\mathcal{F}_\theta$ is exact of order θ. Suppose that T is a bounded* operator $T : \mathcal{F}_\theta(X, Y) \rightarrow Z$, with $\|T\|_{\mathcal{F}_\theta(X,Y) \rightarrow Z} \leq \varphi(\theta)$, then for $f \in X \cap Y$, we have

$$\|Tf\|_Z \leq c \|f\|_X \tau(\frac{\|f\|_Y}{\|f\|_X}) \tag{3.6}$$

where $\tau(s) = \inf_\theta \{s^\theta \varphi(\theta)\}$.

Proof. By Theorem 2, we have, with $\rho(t) = \sup_{\theta \in (0,1)}\{t^\theta \varphi(\theta)^{-1}\}$,

$$\|Tf\|_Z \leq \|f\|_{\sum(\varphi(\theta)\bar{A}_{\theta,1;J})} \approx \|f\|_{\bar{A}_{\rho,1;J}}$$

and if $f \in X \cap Y$, then

$$\|f\|_{\bar{A}_{\rho,1;J}} \leq \inf_{t>0}\{\frac{J(t,f;X,Y)}{\rho(t)}\} = \frac{\|f\|_X}{\rho(\frac{\|f\|_X}{\|f\|_Y})}$$

Thus, if we let

$$\tau(s) = \frac{1}{\rho(\frac{1}{s})} = \inf_\theta\{s^\theta \varphi(\theta)\}$$

then

$$= \|f\|_X \tau(\frac{\|f\|_Y}{\|f\|_X})$$

as desired. $\square$

3.2 Applications: Sobolev Imbeddings

In this section we consider applications to Sobolev imbedding theorems. Let us record first the following general remark

Example 17 *Let $\bar{A}$ be an ordered mutually closed pair, and let X be a Banach space. We consider the functions (cf. the discussion below Theorem 2) $\rho_\alpha(t) \approx (1 + \log e/t)^{-\alpha}$, $\alpha > 0$, $t \in (0, 1)$, i.e. $\rho_\alpha(t) \approx \frac{1}{f_\alpha(\frac{1}{t})}$, where $f_\alpha(t) = \log^\alpha(e^2 + t)$. Then, by Theorem 15, and the characterization of the extrapolation spaces $\bar{A}_{(\alpha)}$, we obtain that for a linear operator $T : A_1 \rightarrow X$,*

$$\|Tx\|_X \leq c \|x\|_{A_0}(1 + \log \frac{\|x\|_{A_1}}{\|x\|_{A_0}})^\alpha \iff T : \bar{A}_{(\alpha)} \rightarrow X$$

In connection with Sobolev spaces let us show a result from [10] (cf. also [57])

Example 18 *There exists an absolute constant $c > 0$ such that $\forall f \in W_2^2(R^2)$, we have*

$$\|f\|_{L^\infty} \leq c(1 + \log(\frac{\|f\|_{W_2^2(R^2)}}{\|f\|_{W_2^1(R^2)}}))^{1/2} \qquad (3.7)$$

Proof. *Observe (cf. [57]) that the embedding*

$$i : [W_2^1(R^2), W_2^2(R^2)]_\theta \to L^\infty(R^n)$$

has norm $\leq c\theta^{-1/2}$, and therefore by extrapolation

$$i : (W_2^1(R^2), W_2^2(R^2))_{(\frac{1}{2})} \to L^\infty(R^2).$$

Now, (3.7) follows from Example 17. $\square$

In our next example we extend a result concerning Calderón-Zygmund operators obtained in [5].

Example 19 *Let T be a Calderón-Zygmund operator, then it is shown in [5] that for $s > n/2$*

$$\|Tx\|_{L^\infty(R^n)} \leq c\|x\|_{L^\infty(R^n)}(1 + \log(\frac{\|x\|_{W_2^s(R^n)}}{\|x\|_{L^\infty(R^n)}}))$$

Note that, by Sobolev's embedding theorem, $(L^\infty(R^n), W_2^s(R^n))$ is an ordered pair. It follows, from Example (17), that we also have

$$T : (L^\infty(R^n), W_2^s(R^n))_{(1)} \to L^\infty(R^n)$$

In [99] it is shown that the result of Example 19 is a simple consequence of the following K/J inequality

Theorem 20 *(cf. [99]) Let $\delta > 0$, and suppose that $s > n/2 + \delta$. Then, there exists an absolute constant $c > 0$ such that $\forall \varepsilon > 0$,*

$$\|f\|_{L^\infty(R^n)} \leq c\varepsilon^\delta\|f\|_{W_2^s(R^n)} + c(\log\frac{1}{\varepsilon})\|f\|_{B_{\infty\infty}^0(R^n)} \qquad (3.8)$$

Proof. The choice $\varepsilon^\delta = \dfrac{\|f\|_{B^0_{\infty\infty}(R^n)}}{\|f\|_{W^s_2(R^n)}}$, transforms (3.8) into

$$\|f\|_{L^\infty(R^n)} \le c\|f\|_{B^0_{\infty\infty}(R^n)}(1 + \log(\|f\|_{W^s_2(R^n)}/\|f\|_{B^0_{\infty\infty}(R^n)})) \quad (3.9)$$

The relationship to Sobolev's embedding theorems can be now be made explicit. Recall that we have (cf. [8])

$$W^s_2(R^n) \subset B^{s-n/2}_{\infty\infty}(R^n)$$

$$B^\theta_{\infty\infty}(R^n) \overset{c\theta^{-1}}{\subset} L^\infty(R^n) \ , \ \theta > 0 \quad (3.10)$$

(cf. [93], Corollary 31).
Then,

$$[B^0_{\infty\infty}(R^n), W^s_2(R^n)]_\theta \subset [B^0_{\infty\infty}(R^n), B^{s-n/2}_{\infty\infty}(R^n)]_\theta = B^{\theta(s-n/2)}_{\infty\infty}(R^n)$$

and by (3.10) we get

$$[B^0_{\infty\infty}(R^n), W^s_2(R^n)]_\theta \overset{c(s)\theta^{-1}}{\subset} L^\infty(R^n)$$

Thus, (3.9) follows from Example 17 and we are done. $\square$

Remark. Using the results of [93] we can also state vector valued analogues. The corresponding results for singular integral operators can be also obtained in this setting once appropriate assumptions are made on the Banach spaces.

Chapter 4

Calculations with the Δ method and applications

Although in previous sections we have focused mainly on computations with the $\sum$ method it is easy to obtain, directly or by duality, suitable analogues for the Δ method. In this section we consider some calculations with the Δ method which are motivated by applications to the theory of orientation preserving maps, and Sobolev imbedding theorems. In fact, our elementary calculations can be used to sharpen recent results by Müller [82] and Iwaniec and Sbordone [54] on the integrability of the Jacobian of orientation preserving maps. We also consider a sharpening of recent results by Fusco, P. L. Lions and Sbordone on limiting cases of Sobolev embedding theorems.

4.1   Reiteration and the Δ method

We shall not attempt to present here the most general results but simply develop the necessary machinery to carry out the calculations in our examples. Thus, the methods are more important than the specific results obtained.

The following easy result, which is the analog of (2.3) for the $\sum$ method, will get us started. Let $\bar{A}$ be an ordered pair, let $\Theta = (0,1)$, let $M(\theta)$ be a positive continuous function such that $M(\theta) \neq 0, \forall \theta \in$

Θ, and for any $\theta_0 \in \Theta$, let $\Theta_0 = (\theta_0, 1)$, then

$$\Delta_{\theta \in \Theta}(M(\theta)\bar{A}_{\theta,q(\theta);K}) = \Delta_{\theta \in \Theta_0}(M(\theta)\bar{A}_{\theta,q(\theta);K}) \qquad (4.1)$$

In fact, we trivially have

$$\sup_{\theta \in \Theta_0} M(\theta)\|a\|_{\bar{A}_{\theta,q(\theta);K}} \leq \sup_{\theta \in \Theta} M(\theta)\|a\|_{\bar{A}_{\theta,q(\theta);K}},$$

while on the other hand,

$$\sup_{\Theta} M(\theta)\|a\|_{\theta,q;K} \leq \sup_{\Theta_0} M(\theta)\|a\|_{\theta,q;K} + \sup_{\Theta_0^c} M(\theta)\|a\|_{\theta,q;K}.$$

Now, since the pair is ordered and $\sup_{\theta \in (0,\theta_0)} M(\theta) = c(\theta_0) < \infty$, we have

$$\sup_{\theta \in \Theta_0^c = (0,\theta_0)} M(\theta)\|a\|_{\theta,q;K} \leq c(\theta_0)M(\theta_0)^{-1} \sup_{\Theta_0} M(\theta)\|a\|_{\theta,q;K}$$

and (4.1) follows.

It is interesting to observe here that we could have derived this result by duality from (2.3). In fact, it is easy to see that for a family of Banach spaces $\{A_\theta\}_{\theta \in \Theta}$, with a common dense subset, we have (cf. [57])

$$\Delta_\theta(A_\theta^*) = (\sum_\theta (A_\theta))^*. \qquad (4.2)$$

In particular, we have

$$\sup_{\Theta}(M(\theta)^{-1}\bar{A}_{\theta,\infty;K}^*) = (\sum_{\theta \in \Theta} M(\theta)\bar{A}_{\theta,1;J})^*, \qquad (4.3)$$

where $\bar{A}^*$ denotes the dual pair (A_0, A_1).

Thus, we could use (4.3) and the duality theory of the real method of interpolation to establish many results for the Δ method from the corresponding ones for the $\sum$ method. One drawback of this method is that it may require some extra assumptions on the spaces.

Let us now give a direct proof of the following analog of Theorem 5.

Theorem 21 *Let $\bar{A}$ be a mutually closed pair, and let $M(\theta)$ be a tempered positive function on $\Theta = (0,1)$. Then for any family of interpolation functors $\{\mathcal{F}_\theta\}_{\theta \in \Theta}$, with $\mathcal{F}_\theta$ exact of order θ, we have:*

$$\sup_{\Theta}(M(\theta)^{-1}\mathcal{F}_\theta(\bar{A})) = \sup_{\Theta}(M(\theta)^{-1}\bar{A}_{\theta,\infty;K})$$

Proof. The assumption that the pair $\bar{A}$ is mutually closed is required since we shall use the **SFL** (i.e. the fundamental lemma in its strong form). To be precise we require the following consequence of SFL:

$$\bar{A}_{\theta,1;K} \subset \bar{A}_{\theta,1;J} \tag{4.4}$$

with norm independent of $\theta \in \Theta$.

To prove (4.4) proceed as follows. Let $a = \int_0^\infty u(s)\frac{ds}{s}$, with

$$\int_0^\infty \min\{1, \frac{t}{s}\} J(s, u(s))\frac{ds}{s} \leq c\, K(t, a).$$

Then,

$$\|a\|_{\bar{A}_{\theta,1;J}} \leq \int_0^\infty s^{-\theta} J(s, u(s))\frac{ds}{s}$$

$$\leq (1-\theta)\theta \int_0^\infty J(s, u(s)) \int_0^\infty \min\{1, \frac{t}{s}\} t^{-\theta}\frac{dt}{t}\frac{ds}{s}$$

$$= (1-\theta)\theta \int_0^\infty \int_0^\infty J(s, u(s)) \min\{1, \frac{t}{s}\}\frac{ds}{s} t^{-\theta}\frac{dt}{t}$$

$$\leq c(1-\theta)\theta \int_0^\infty K(t, a; \bar{A}) t^{-\theta}\frac{dt}{t}$$

$$= c\|a\|_{\bar{A}_{\theta,1;K}}$$

as required.

Let us also recall from Chapter **2** the fact that if $M(\theta)$ is tempered then, with $\tau(t) = \inf_\theta M(\theta)t^\theta$, we have

$$M(\theta) \approx [(1-\theta)\theta] \int_0^\infty t^{-\theta}\tau(t)\frac{dt}{t}. \tag{4.5}$$

With these auxiliary results at hand we may now complete the proof of the theorem. It is clearly sufficient to prove that

$$\sup_\theta (M(\theta))^{-1}\|a\|_{\bar{A}_{\theta,1;K}\theta} \leq c \sup_\theta (M(\theta))^{-1}\|a\|_{\bar{A}_{\theta,\infty;K}}$$

Now,

$$\sup_\theta (M(\theta))^{-1}\|a\|_{\bar{A}_{\theta,1;K}} \leq c \sup_\theta (M(\theta))^{-1}(1-\theta)\theta \int_0^\infty K(s, a)s^{-\theta}\frac{ds}{s}$$

$$\leq c \sup_{\theta}[(1-\theta)\theta](M(\theta))^{-1} \int_0^\infty \tau(s)s^{-\theta}\frac{ds}{s} \sup_{t>0}\left(\frac{K(t,a)}{\tau(t)}\right)$$

$$\leq c\sup_{t>0}\left(\frac{K(t,a)}{\tau(t)}\right)\ \ (\text{by } (4.5))$$

$$= c \sup_{t>0}\ K(t,a)[\frac{1}{inf_\theta\{M(\theta)t^\theta\}}]$$

$$= c\sup_{t>0}\ K(t,a)\sup_{\theta}\{(M(\theta))^{-1}\ t^{-\theta}\}$$

$$= c\sup_{\theta}\ (M(\theta))^{-1}\sup_{t>0}\ K(t,a)t^{-\theta}$$

$$= c\ \sup_{\theta}(M(\theta))^{-1}\|a\|_{\bar{A}_{\theta,\infty;K}}$$

and the result follows. □

Let us remark that the same calculation proves that

$$\|a\|_{\Delta(t^\theta M(\theta)^{-1}\mathcal{F}_\theta(\bar{A}))} \approx \sup_{s>0}[\frac{K(s,a;\bar{A})}{\tau(\frac{s}{t})}] \tag{4.6}$$

where $\tau(t) = \inf\{M(\theta)t^\theta\}$.

Example 22 *As we have pointed out in (2.27) it is easy to see that*

$$\Delta_{1<p<\infty}(p^\alpha L(p,\infty)) = Exp\ L^{1/\alpha}$$

Thus, it follows from the Theorem 21 that

$$\Delta_{1<p<\infty}(p^\alpha L^p) = Exp\ L^{1/\alpha}$$

4.2   On the Integrability of Orientation Preserving Maps

Recently it has been discovered that the Jacobians of orientation preserving maps, and other related nonlinear quantities, enjoy better integrability properties than those known for the Jacobians of standard maps. The first results in this direction were obtained by Müller [82], and have been extended in many different directions by a number of authors including Coifman, Lions, Meyer, and Semmes

[19], Iwaniec and Sbordone [54], Brezis, Fusco and Sbordone [9], Iwaniec and Lutoborski [53], Iwaniec and Greco [44], and many others. These developments have interesting applications to the study of the equations of non-linear elasticity, variational problems, compensated compactness, etc. We must refer the reader to these papers for a detailed and complete treatment. In Chapter **7** we shall give a more detailed account of applications of commutator methods to compensated compactness. In particular, there we prove some higher integrability theorems for Jacobians.

In this section we show the relevance of the Δ method in this theory.

4.2.1 Background

Let Ω be a bounded open set in R^n, $f : \Omega \to R^n$ be a *smooth* mapping, we say that f is *orientation preserving* if its Jacobian $Jf = \det Df$ is nonnegative a.e.. A typical assumption on the smoothness of f is that f is in the Sobolev class $W_n^1(\Omega, R^n)$. Observe that by Hadamard's inequality $Jf \leq |\nabla f_1| \cdots |\nabla f_n|$, and therefore by Hölder's inequality we have that $f \in W_n^1(\Omega, R^n)$ implies that $Jf \in L^1(\Omega)$. It was recently discovered by Müller [82] that if f is an *orientation preserving* map then one has a better result

Theorem 23 *Let Ω be a bounded open domain in $R^n, n \geq 2$, and let $f : \Omega \to R^n$ be an orientation preserving map in the Sobolev class $W_n^1(\Omega, R^n)$. Then $\forall K \subset \Omega$, compact we have that $Jf \in L(LogL)(K)$.*

This result has now been extended and applied in many different directions by a number of authors. In [19] Coifman, Lions, Meyer and Semmes point out the rôle of the Hardy space $H^1(R^n)$ proving, among other things, the following

Theorem 24 *Let $f : R^n \to R^n$, $f \in W^1(R^n, R^n)$, then $Jf \in H^1(R^n)$.*

The relationship with Müller's result is given by a theorem of Stein (cf. [98]) stating that if $g \geq 0, then\ g \in H^1_{loc}(R^n) \iff g \in L(LogL)_{loc}(R^n)$.

Remark. In Chapter **7**, Theorem 78, we give a sharpening of Theorem 24, for smooth maps, using commutator techniques.

In a different direction Iwaniec and Sbordone give in their paper [54] sufficient conditions to guarantee the *local integrability* of the Jacobian of an orientation preserving map

Theorem 25 *Let $B \subset 3B$ be concentric balls in R^n, and let f : $3B \to R^n$ be an orientation preserving map. Then,*

$$f \in \Delta_{s\in[1,n)}(n-s)W_s^1(3B, R^n) \Longrightarrow Jf \in L^1(B).$$

Here we denote by $\Delta_{s\in[1,n)}(n-s)W_s^1(3B, R^n)$ the space of maps such that

$$\sup_{s\in[1,n)} (n-s)\|f\|_{W_s^1(3B,R^n)} < \infty$$

It is shown in [54] that

$$W^1_{L^n(LogL)^{-1}}(3B, R^n) \subset \Delta_{s\in[1,n)}(n-s)W_s^1(3B, R^n) \qquad (4.7)$$

$$W^1_{L^n,\infty}(3B, R^n) \subset \Delta_{s\in[1,n)}(n-s)W_s^1(3B, R^n) \qquad (4.8)$$

where for a function space X, and a domain Ω, we denote by $W_X^1(\Omega, R^n)$ the class of maps $f = (f_1, ... f_n)$ with components such that $\nabla f_i \in X(3B)$, $i = 1, ... n$.

Corollary 26 *Let $B \subset 3B$ be concentric balls in R^n, and let f : $3B \to R^n$ be an orientation preserving mapping, $f \in W^1_{L^n(LogL)^{-1}}(3B, R^n) \cup W^1_{L^n,\infty}(3B, R^n)$, then, $Jf \in L^1(B)$.*

More recently, the author obtained the following (cf. [74], [75])

Theorem 27 *Let Ω be a bounded open domain of R^n, and let f be an orientation preserving map of class $W^1_{L^n(LogL)^\theta}(\Omega, R^n)$ then $Jf \in L^n(LogL)^{\theta+1}(K)$, $\forall K \subset \Omega, \forall \theta \in R$.*

Remark. The case $\theta = -1$ corresponds to the result of Iwaniec-Sbordone, the case $\theta \in [-1,0]$ is due to Brezis, Fusco and Sbordone [9], while the case $\theta = 1$ is due to Greco and Iwaniec [44].

Although we cannot go into the details here we would like to at least illustrate the import of these estimates in the study of the weak compactness of the Jacobian map. For example, as an application of his theory, Müller [82] proves that if $\{u_j\}_{j \in N}$ is a sequence of orientation preserving mappings, $u_j : \Omega \to R^n$, and $u_j \rightharpoonup u$ (weakly) in $W_n^1(\Omega, R^n)$, then $\forall K \subset \Omega$ compact, we have

$$J(x, u_j) \rightharpoonup J(x, u) \text{ weakly in } L^1(K) \tag{4.9}$$

To see the connection with the $LLogL$ estimates we recall the classical criteria of de La Vallée Poussin stating that for a set K of finite measure, a sequence $\{f_j\}_{j \in N}$ is relatively weakly sequentially compact in $L^1(K)$ if and only if there exists a positive function γ defined on R_+ with $\lim_{x \to \infty} \gamma(x)/x = \infty$ such that

$$\sup_j \int_K \gamma(|f_j(x)|) dx < \infty \tag{4.10}$$

Now, it was known (cf. [4]) that if $u_j \rightharpoonup u$ (weakly) in $W_n^1(\Omega, R^n)$ then

$$J(x, u_j) \rightharpoonup^* J(x, u) \text{ weak}^*$$

in the sense of measures. Therefore, under the extra assumption that the maps are orientation preserving, we may apply Theorem 23 and de La Vallée Pousin's criteria to obtain the stronger conclusion (4.9). Theorem 27 combined with a general form of de La Vallée Pousin's criteria can be used to obtain similar convergence results for the Jacobians of orientation preserving maps in the $W_{L^n(LogL)^\theta}^1(\Omega, R^n)$ classes (cf. [79]).

4.2.2 Identification of Sobolev Classes using Δ

In this section we give a characterization of the spaces that appear in Theorem 25. For the benefit of the reader we shall give a complete approach to the relevant part of the interpolation theory of Sobolev spaces.

Working with the coordinate components we shall reduce ourselves to work with ordinary Sobolev spaces. Thus, we consider the Sobolev spaces

$$W_p^k(\Omega) = W_p^k(\Omega) = \{f : D^\alpha f \in L^p(\Omega), |\alpha| \le k\}$$

where Ω is a regular domain (which for our purposes we shall take to mean that all functions in $W_p^k(\Omega)$ admit extensions to functions in $W_p^k(R^n)$, i.e. the existence of an extension operator), $k \geq 0$, $1 \leq p \leq \infty$, and the derivatives are taken in the sense of distributions. The corresponding norms are given by

$$\|f\|_{W_p^k} = \sum_{|\alpha| \leq k} \|D^\alpha f\|_p.$$

The following result is due to DeVore and Scherer [34], but we shall present here the proof given by C. P. Calderón and the author in [15].

Theorem 28 *If Ω is a regular domain, then*

$$K(t, f, W_1^k(\Omega), W_\infty^k(\Omega)) \approx \sum_{|\alpha| \leq k} K(t, D^\alpha f, L^1(\Omega), L^\infty(\Omega))$$

$$\approx \sum_{|\alpha| \leq k} \int_0^t (D^\alpha f)^*(s) ds$$

In order to carry out the proof we need to develop certain preliminary material. We start by developing a technique to estimate remainders of Taylor expansions in terms of Maximal operators. This technique was initially developed by Calderón and Zygmund in the sixties but the results we need here were apparently first derived in [15].

Let $f \in C_0^k(R^n)$, and given $x, y \in R^n$, consider the Taylor expansions

$$D^\alpha f(x) = \sum_{|\alpha| \leq k-1} D^{(\alpha+l)} f(y) \frac{(x-y)^l}{l!} + R_\alpha(x, y)$$

For $R_\alpha(x, y)$ we can use the alternate expressions

$$R_\alpha(x, y) = (k - |\alpha|) \sum_{|\alpha+l|=k} \frac{(x-y)^l}{l!} \int_0^1 t^{k-|\alpha|-1} D^{(\alpha+l)} f(x + t(y - x)) dt$$

or

$$\hat{R}_\alpha(x, y) = (k - |\alpha|) \sum_{|\alpha+l|=k} \frac{(x-y)^l}{l!} \int_0^1 (1-t)^{k-|\alpha|-1} D^{(\alpha+l)} f(y + t(x - y)) dt$$

Lemma 29 *There exists an absolute constant $C > 0$, such that*

$$\sup_{\delta > 0} \delta^{-n} \int_{|x-y|<\delta} \delta^{|\alpha|-k} |R_\alpha(x,y)| dy \leq C \sum_{|\alpha+l|=k} M(D^{(\alpha+l)}f)(x)$$

$$\sup_{\delta > 0} \delta^{-n} \int_{|x-y|<\delta} \delta^{|\alpha|-k} |\hat{R}_\alpha(x,y)| dx \leq C \sum_{|\alpha+l|=k} M(D^{(\alpha+l)}f)(y)$$

where M is the maximal operator of Hardy and Littlewood.

Proof.   We prove only the first estimate. By a translation and a change of scale, one can reduce to the case $\delta = 1$, $x = 0$. Then we see that everything is a consequence of the following fact: if g is a positive measurable function and we set $h(x) = \int_0^1 g(tx)dt$. Then,

$$\int_{|x| \leq 1} h(x)dx \leq c \, Mg(0)$$

To prove this we use Fubini's theorem and a change of variable ($tx = u$) to derive

$$\int_{|x| \leq 1} h(x)dx = \int_0^1 \int_{|x| \leq 1} g(tx)dx \, dt$$

$$= \int_0^1 t^{-n} \int_{|x| \leq t} g(u)du dt$$

$$\leq c \, Mg(0).$$

$\square$

   Proof of Theorem 28. Since Ω is regular we may assume that $\Omega = R^n$. The estimate

$$\sum_{|\alpha| \leq k} K(t, D^\alpha f, L^1, L^\infty) \leq cK(t, f, W_1^k, W_\infty^k)$$

is obvious. To prove the converse we may, by an approximation argument, restrict ourselves to smooth functions. Let $f \in C_0^k$, and let $t > 0$ be given. For each $\alpha \in Z_+^n$, with $|\alpha| \leq k$, let

$$E_\alpha = \{x : M(D^\alpha f)(x) > [M(D^\alpha f)]^*(t)\}$$

and $E = \bigcup_{|\alpha| \le k} E_\alpha$. Then, E is open and $|E| \le ct$.

Let $f^- = f \; \chi_{E^c}$, then we shall prove in a moment that $f^- \in Lip(k, E^c)$, and, moreover, that its norm in this space is less than or equal than a constant times $\sum_{\alpha| \le k} [M(D^\alpha f)]^*(t)$. At this point it follows, by the Whitney extension theorem (cf. [98]), that f^- admits an extension $f_\infty \in W^k_\infty$, such that

$$\|f_\infty\|_{W^k_\infty} \le c \sum_{|\alpha| \le k} [M(D^\alpha f)]^*(t).$$

Combining this estimate with the well known facts that

$$[M(h)]^*(t) \le ch^{**}(t), \text{ and } K(t, h, L^1, L^\infty) = th^{**}(t),$$

we get

$$t \|f_\infty\|_{W^k_\infty} \le c \sum_{|\alpha| \le k} K(t, D^\alpha f, L^1, L^\infty) \qquad (4.11)$$

Define $f_1 = f - f_\infty$, then

$$\|f_1\|_{W^k_1} \le \sum_{|\alpha| \le k} \|D^\alpha f \; \chi_E\|_{L^1} + \sum_{|\alpha| \le k} \|D^\alpha f_\infty \; \chi_E\|_{L^1} = I_1 + I_2.$$

To estimate each term in the sum I_1, we observe that

$$\|D^\alpha f \; \chi_E\|_{L^1} \le \int_0^{ct} (D^\alpha f)^*(s) ds$$

$$\le c \int_0^t (D^\alpha f)^*(s) ds \text{ (since } (D^\alpha f)^{**}(t) \text{ is decreasing)}$$

$$= cK(t, D^\alpha f, L^1, L^\infty)$$

Having obtained the correct estimate for I_1 we turn to I_2. Again consider a typical term of the sum I_2, then

$$\|D^\alpha f_\infty \; \chi_E\|_{L^1} \le ct[M(D^\alpha f)]^*(t) \le c \; K(t, D^\alpha f, L^1, L^\infty)$$

and therefore, collecting these estimates, we have obtained

$$\|f_1\|_{W^k_1} \le cK(t, D^\alpha f, L^1, L^\infty) \qquad (4.12)$$

In view of (4.11) and (4.12), we see that we have constructed a decomposition $f = f_1 + f_\infty$, with $f_1 \in W^k_1$, $f_\infty \in W^k_\infty$, such that

$$K(t, f; W^k_1, W^k_\infty) \le \|f_1\|_{W^k_1} + t \|f_\infty\|_{W^k_\infty}$$

$$\leq cK(t, D^\alpha f, L^1, L^\infty).$$

The theorem is proved modulo proving our assertion about the function f^-. Recall that we are required to estimate the Lip norm of f^- in E^c. We start with the higher order remainders. Let $\alpha \in Z_+^n$, with $|\alpha| = k - 1$, and let $x_1, x_2 \in E^c$, we estimate $D^\alpha f(x_1) - D^\alpha f(x_2)$. To do so let $r = |x_1 - x_2|$, and let Q_i, be cubes centered at x_i, $i = 0$, 1, with sides parallel to the coordinate axes with length equal to $\frac{4}{3} r$. Then, by simple geometry, there exists a constant c_n independent of the $x_i's$ such that $|Q_1 \cap Q_2| > c_n |Q_i|$, $i = 0, 1$. Let

$$H_i = \{y \in Q_i : \frac{|D^\alpha f(x_i) - D^\alpha f(y)|}{r} > \lambda(t)N\},$$

where $\lambda(t) = \sum_{|\beta| \leq k} [M(D^\beta f)]^*(t)$, and N is a positive fixed number to be determined precisely later on. We shall show that it is possible to choose N so large that

$$|H_1 \cup H_2| < |Q_1 \cap Q_2|.$$

Thus, it is possible to find points in the set $Q_1 \cap Q_2 \setminus H_1 \cup H_2$. Let z be one such point, then we can estimate

$$|D^\alpha f(x_i) - D^\alpha f(z)| \leq r \, \lambda(t)N, \, i = 0, 1$$

and therefore we derive the desired estimate

$$|D^\alpha f(x_0) - D^\alpha f(x_1)| \leq |x_0 - x_1| \lambda(t)N$$

$\forall x_0, x_1 \in E$.

The lower order remainders can be now estimated using the formulae (cf. [98])

$$|R_\alpha(x_1, x_2)| \leq |R_\alpha(x_1, z)| + \sum_{|\alpha + l| \leq k - 1} |R_{\alpha + l}(z, , x_2)| \frac{|x_1 - z|^l}{l!}.$$

and we obtain the required

$$|R_\alpha(x_1, x_2)| \leq C |x_0 - x_1| \lambda(t)$$

It remains, thus, to prove the assertion about the measure of the union of the sets H_i. It is at this point that the maximal operators of the previous lemma appear. In fact, to estimate the measure of H_i we use successively Chebyshev's inequality and the lemma to obtain

$$|H_i| \leq [\lambda(t)N]^{-1} \int_{Q_i} \frac{|D^\alpha f(x_i) - D^\alpha f(y)|}{r} dy$$

$$\leq C[\lambda(t)N]^{-1} \sum_{|\alpha+l|=k} M(D^{(\alpha+l)}f)(x_i)|Q_i|$$

$$\leq C \; N^{-1}|Q_i| \text{ (since } x_i \in E^c)$$

Therefore, if we choose N so large that $\frac{C}{N} < \frac{c_n}{2}$, we accomplish our goal of showing that

$$|H_1 \cup H_2| < |Q_1 \cap Q_2|.$$

concluding the proof. □

Theorem 30 *If Ω is a regular domain, $1 \leq p_1 < p_2 \leq \infty$, $\frac{1}{\alpha} = \frac{1}{p_1} - \frac{1}{p_2}$, then*

$$K(t, f, W_{p_1}^k(\Omega), W_{p_2}^k(\Omega)) \approx \sum_{|\alpha| \leq k} K(t, D^\alpha f, L^{p_1}(\Omega), L^{p_2}(\Omega))$$

$$\approx \sum_{|\alpha| \leq k} \left\{ \int_0^{t^\alpha} [(D^\alpha f)^*(s)]^{p_1} ds \right\}^{\frac{1}{p_1}} + t \left\{ \int_{t^\alpha}^{|\Omega|} [(D^\alpha f)^*(s)]^{p_2} ds \right\}^{\frac{1}{p_2}}$$

(4.13)

Proof.   It follows from Theorem 28 that

$$\left(W_1^k(\Omega), W_\infty^k(\Omega) \right)_{\frac{1}{p'}, p; K} = W_p^k(\Omega) \tag{4.14}$$

and therefore the result follows by Holmstedt's formula (cf. [8]). □

We consider now the identification of the class of Sobolev maps that is described in Theorem 25.

Theorem 31 $u \in \Delta_{s \in [1,n)}(n - s)W_s^1(3B, R^n)$ *if an only if the distributional derivatives of its components, $\frac{\partial u_i}{\partial x_i}$, are such that*

$$\sup_{t \in (0,1)} \rho(t)[\sum_{i,j} \int_0^{t^{\frac{n}{n-1}}} \left(\frac{\partial u_j}{\partial x_i}\right)^*(s)ds + t\{\int_{t^{\frac{n}{n-1}}}^1 \left[\left(\frac{\partial u_j}{\partial x_i}\right)^*(s)\right]^n ds\}^{\frac{1}{n}}] < \infty$$

(4.15)

where $\rho(t) = t^{-1}(1 + \log \frac{1}{t})^{-1}$

Proof. Observe that by (4.1)

$$\Delta_{s \in [1,n)}(n-s)W_s^1(3B, R^n) = \Delta_{s \in [\frac{3}{2},n)}(n-s)W_s^1(3B, R^n).$$

Now, by (4.14), reiteration, and Theorem 21 we see that

$$\Delta_{s \in [\frac{3}{2},n)}(n-s)W_s^1(3B, R^n)$$

$$= \Delta_{\theta > \frac{1}{3}}(1-\theta)(W_1^1(3B, R^n), W_n^1(3B, R^n))_{\theta, \infty; K}$$

By (4.13) we have

$$K(t, f, W_1^1(3B, R^n), W_n^1(3B, R^n)) \approx$$

$$\approx [\sum_{i,j} \int_0^{t^{\frac{n}{n-1}}} \left(\frac{\partial u_j}{\partial x_i}\right)^*(s)ds + t\{\int_{t^{\frac{n}{n-1}}}^1 \left[\left(\frac{\partial u_j}{\partial x_i}\right)^*(s)\right]^n ds\}^{\frac{1}{n}}]$$

Therefore the desired characterization follows from Theorem 2. $\square$

Using Theorem 31 it is easy to give a proof of (4.7) and (4.8). Let us consider, for example, (4.7). Suppose then that $u \in W_{L^n,\infty}^1(\Omega, R^n)$, that is

$$\|u\|_{W_{L^n,\infty}^1(\Omega,R^n)} \approx \sup_s s^{\frac{1}{n}} \sum_{i,j} \left(\frac{\partial u_j}{\partial x_i}\right)^*(s) < \infty.$$

We must estimate each of the terms in (4.15):

$$\rho(t)t\{\int_{t^{\frac{n}{n-1}}}^1 \left[\left(\frac{\partial u_j}{\partial x_i}\right)^*(s)\right]^n ds\}^{\frac{1}{n}} \leq ct^{-1}(1 + \log \frac{1}{t})^{-1}t[\log \frac{1}{t}]^{1/n} \leq c$$

and

$$\rho(t)\int_0^{t^{\frac{n}{n-1}}} \left(\frac{\partial u_j}{\partial x_i}\right)^*(s)ds \leq c\, t^{-1}(1 + \log \frac{1}{t})^{-1}t \leq c$$

the result follows by Theorem 31.

4.3   Some Extreme Sobolev Imbedding Theorems

This section was motivated by recent work by N. Fusco-P. L. Lions-C. Sbordone [42] on extreme forms of Sobolev imbedding theorems. The main result in [42] is the following

Theorem 32 *Let Ω be an open domain in R^n with $|\Omega| < \infty$. Let $u \in W_0^{1,1}(\Omega)$ satisfy for some $\sigma \geq 0$*

$$\|u\|_\sigma = \sup_{0<\varepsilon\leq 1} \left(\varepsilon^\sigma \, |\Omega|^{-1} \int_\Omega |Du|^{n-\varepsilon} \, dx \right)^{\frac{1}{n-\varepsilon}} < \infty. \qquad (4.16)$$

Then, if $\alpha = \frac{n}{n-1+\sigma}$, there exist constants $c_1(n,\sigma), c_2(n,\sigma)$ such that

$$\int_\Omega \exp(\frac{|u|}{c_1 M \, |\Omega|^{\frac{1}{n}}})^\alpha dx \leq c_2.$$

In this section we discuss the relationship of this result with extrapolation. First, the reader should have no problems in verifying that

$$\|u\|_\sigma < \infty \Leftrightarrow u \in \Delta_\theta(1-\theta)^{\frac{n-1+\sigma}{n}} \left(W_1^1(\Omega), W_n^1(\Omega) \right)_{\theta,\infty;K} \qquad (4.17)$$

Let $M_\gamma = M_\gamma(\Omega)$ denote the space defined by the condition $f \in M_\gamma$ if and only if

$$\sup_{t\in(0,1)} \rho_\gamma(t)[\sum_i \int_0^{t^{\frac{n}{n-1}}} f^*(s)ds + t\{\int_{t^{\frac{n}{n-1}}}^1 (f^*(s))^n \, ds\}^{\frac{1}{n}}] < \infty \qquad (4.18)$$

where $\rho_\gamma(t) = t^{-1}(1 + \log \frac{1}{t})^{-\gamma}$. Then, using Theorem 31 (or more precisely its proof) we find that (4.17) is equivalent to the condition $\frac{\partial u}{\partial x_i} \in M_{\frac{n-1+\sigma}{n}}, i = 1..,n$.

Let us now review the proof of Theorem 32 following [42]. We start with the known estimates for the potential operator

$$If(x) = \int_\Omega \frac{f(y)}{|x-y|^{n-1}} dy$$

$$\|If\|_{L^q} \leq \left(\frac{1-\delta}{\frac{1}{n}-\delta}\right)^{1-\delta} \omega_n^{1-\frac{1}{n}} |\Omega|^{\frac{1}{n}-\delta} \|f\|_{L^p} \qquad (4.19)$$

where $0 \leq \delta = \frac{1}{p} - \frac{1}{q} < \frac{1}{n}$.

It is easy to see that (4.19) implies that

$$I : (L^1, L^n)_{\theta, p(\theta); K} \rightarrow (1-\theta)^{\frac{n-1}{n}} c_n (L^1, L^\infty)_{\theta, q(\theta); K} \qquad (4.20)$$

Thus, applying the Δ method, we arrive to

$$I : \Delta_\theta (L^1, L^n)_{\theta, p(\theta); K} \rightarrow \Delta_\theta (1-\theta)^{\frac{n-1}{n}} c_n (L^1, L^\infty)_{\theta, q(\theta); K}$$

which gives :

$$I : L^n \rightarrow e^{L^{\frac{n}{n-1}}}$$

If we *multiply*(!) (4.20) by $(1-\theta)^{\frac{\sigma}{n}}$ and *then* apply Δ we get:

$$I : \Delta_\theta (1-\theta)^{\frac{\sigma}{n}} (L^1, L^n)_{\theta, p(\theta); K} \rightarrow \Delta_\theta (1-\theta)^{\frac{n-1}{n} + \frac{\sigma}{n}} c_n (L^1, L^\infty)_{\theta, q(\theta); K}$$

Therefore we get

$$I : M_{\frac{\sigma}{n}} \rightarrow e^{L^{\frac{n}{n-1+\sigma}}}$$

Using the fact that

$$|u(x)| \leq I(|Du|)(x)$$

the discussion above gives the Sobolev imbedding theorem

$$W^1_{M_{\frac{\sigma}{n}}}(\Omega) \subset e^{L^{\frac{n}{n-1+\sigma}}}$$

Note that, when $\sigma = 0$,

$$W^1_{M_{\frac{\sigma}{n}}}(\Omega) = W^1_n(\Omega)$$

and we obtain Trudinger's Sobolev imbedding theorem.

Chapter 5

Bilinear Extrapolation And A Limiting Case of a Theorem by Cwikel

In this chapter we discuss bilinear extrapolation. We provide a general approach to the extrapolation of bilinear operators including an extension of the classical extrapolation theorem of Yano. We also illustrate the use of bilinear extrapolation as a tool in analysis. In fact, in our main application in this chapter, we treat in detail certain bilinear operators, with values on Schatten ideals, studied by Cwikel [28]. We show that we can extrapolate from Cwikel's Theorem certain estimates derived by Constantin [22]. In turn the results of [22] have interesting applications: they provide an end point to the so called collective Sobolev imbedding theorems (cf. [70], [100]) and are useful in the theory of the Navier Stokes equations.

In this chapter we also show that the Macaev ideals are extrapolation spaces associated to the Schatten S_p scale. In fact we show, following [58], that the Macaev ideals are the counterpart in the setting of operator ideals of the $LLogL$ and e^L spaces in the L^p scale. Using this insight it is possible to formulate extensions of Cwikel's theorem in terms of the Macaev ideals and some variants of them.

5.1 Bilinear Extrapolation

We start our program developing in more detail some ideas already outlined in [58] for bilinear extrapolation in the setting of L^p spaces. For conciseness sake we shall refer to [57], and [58] for proofs of the statements, whenever possible.

Theorem 33 *(cf.[58]) Let $\bar{A}$, $\bar{B}$, $\bar{C}$, be Banach pairs, and let T be a bilinear operator, defined on $\Delta(\bar{A}) \times \Delta(\bar{B})$ with values on a pair $\bar{C}$. Let $M(\theta)$ be a positive function of θ, $\theta \in (0,1)$, $\tau(t) = \inf_\theta M(\theta)t^\theta$. Then, the following two conditions are equivalent for T :*

i) $T : \bar{A}_{\theta,1;J} \times \bar{B}_{\theta,1;J} \overset{M(\theta)}{\to} \bar{C}_{\theta,\infty;K}, 0 < \theta < 1$,
ii) $K(t, T(a, b); \bar{C}) \leq c\inf \int_0^\infty \int_0^\infty J(s, u(s); \bar{A})J(r, v(r); \bar{B})\tau(\frac{t}{sr})\frac{ds}{s}\frac{dr}{r}$, where the infimum is taken over all possible decompositions $a = \int_0^\infty u(s)\frac{ds}{s}$, $b = \int_0^\infty v(r)\frac{dr}{r}$.

In what follows we need to impose more restrictions on the growth of the function $M(\theta)$. Consider the function $\tau(t) = \inf_\theta t^\theta M(\theta)$, then, if τ is twice differentiable, we can write

$$\tau(t) = \int_0^\infty \min(1, \frac{t}{r})(-r^2\tau''(r))\frac{dr}{r}.$$

Now, suppose moreover that $(-r^2\tau''(r))$ is itself a quasi-concave function, with

$$\lim_{t\to 0} -r^2\tau''(r) = \lim_{t\to\infty} \frac{-r^2\tau''(r)}{t} = 0.$$

Then, there exists a measure $d\mu$ such that

$$\tau(t) = \int_0^\infty \min(1, \frac{t}{r}) \int_0^\infty \min(1, \frac{r}{s})d\mu(s)\frac{dr}{r}$$

More generally, let us say that $M(\theta)$ is *presentable* if the associated τ admits a representation of the form:

$$\tau(t) = \int_0^\infty \int_0^\infty \min(1, \frac{t}{s}) \min(1, \frac{s}{r})d\mu(r)\frac{ds}{s} \qquad (5.1)$$

for some positive measure $d\mu$.

Let us observe that (5.1) can be equivalently rewritten as

$$= \int_0^\infty \int_0^\infty \min(1, \frac{t}{r})(2 + \log \frac{t}{r})d\mu(r)\frac{ds}{s} \qquad (5.2)$$

Under this assumption, and through the use of SFL we can prove that the conditions of Theorem 33 can be sharpened as follows

Theorem 34 *(cf. [58]) Suppose that the assumptions of Theorem 33 hold and moreover, $\bar{A}, \bar{B}$ are mutually closed pairs, and $M(\theta)$ is presentable. Then, all the conditions of Theorem 33 are equivalent to*

$$K(t, T(a, b); \bar{C}) \le c \int_0^\infty \int_0^\infty K(s, a; \bar{A})K(\frac{t}{sr}, b; \bar{B})\frac{ds}{s}d\mu(r),$$

$\forall t > 0$, $a \in \Delta(\bar{A})$, $b \in \Delta(\bar{B})$, *for some universal constant $c > 0$.*

Proof.   Let $a = \int_0^\infty u(s)\frac{ds}{s}$, $b = \int_0^\infty v(u)\frac{du}{u}$, be representations provided by the SFL, then, by a familiar argument, we get

$$K(t, T(a, b); \bar{C}) \le c \int_0^\infty \int_0^\infty J(s, u(s); \bar{A})J(u, v(u); \bar{B})\tau(\frac{t}{us})\frac{ds}{s}\frac{du}{u}.$$

Inserting the representation of τ in this formula, and making a change of variables, shows that the right hand side is equal to

$$\int_0^\infty \int_0^\infty J(s, u(s); \bar{A})J(u, v(u); \bar{B}) \int_0^\infty \int_0^\infty \min(1, \frac{t}{ux})\min(1, \frac{x}{sy})d\mu(y)\frac{dx}{x}\frac{dy}{y}\frac{du}{u}\frac{ds}{s}$$

Now, interchanging the order of integration and using the special properties of the representations chosen for a, and b, we get that the last expression is equal to

$$\int_0^\infty \int_0^\infty \int_0^\infty J(s, u(s); \bar{A}) \int_0^\infty J(u, v(u); \bar{B}) \min(1, \frac{t}{ux})\frac{du}{u} \min(1, \frac{x}{sy})d\mu(y)\frac{dx}{x}\frac{ds}{s}$$

$$\leq c \int_0^\infty \int_0^\infty \int_0^\infty J(s, u(s); \bar{A}) \min(1, \frac{x}{sy}) \frac{ds}{s} K(\frac{t}{x}, b; \bar{B}) d\mu(y) \frac{dx}{x}$$

$$\leq c \int_0^\infty \int_0^\infty K(\frac{x}{y}, a; \bar{A}) K(\frac{t}{x}, b; \bar{B}) d\mu(y) \frac{dx}{x}$$

$$= c \int_0^\infty \int_0^\infty K(x, a; \bar{A}) K(\frac{t}{xy}, b; \bar{B}) d\mu(y) \frac{dx}{x}$$

As we wished to show. $\square$

Example 35 *The previous discussion gives us a criteria for deciding which growth functions $M(\theta)$ are presentable. For example, according to Example 3, if $M(\theta) \approx \theta^{-\alpha}, \alpha \geq 1$, then*

$$\tau(t) \approx \begin{cases} \log^\alpha t & t \geq 1 \\ t & t < 1 \end{cases}$$

so that we have

$$-t^2 \tau''(t) \approx \begin{cases} \alpha \log^{\alpha-1} t - \alpha(\alpha-1) \log^{\alpha-2} t & t > 1 \\ 0 & t < 1 \end{cases}$$

and we see that $-t^2\tau''(t)$ is quasi-concave. Therefore, $M(\theta)$ is presentable in this case.

Example 36 *A minor variation of the previous example is given by growth functions of the form $M(\theta) \approx \theta^{-\alpha}(1-\theta)^{-\beta}, \alpha, \beta \geq 1$. We claim that these growth functions are presentable. Let us consider the case where $\alpha = \beta$ in detail. Let $M_k(\theta) = \theta^{-k}(1-\theta)^{-k}, k = 1, ...,$ and let $\tau_k(t)$ be their associated functions defined by $\tau_k(t) = \inf_\theta t^\theta M_k(\theta)$. Then,*

$$\tau_k(t) \approx \min(1, t)(1 + |\log t|)^k. \qquad (5.3)$$

In fact, in a less precise fashion, (5.3) has already been indicated in Example 3. For a proof, we check, by computation, that (5.3) actually holds for $k = 1$. Then, we see that the estimate is actually valid for all k, since $\tau_k(t) \approx \left(\tau_1(t^{\frac{1}{k}})\right)^k$. In the case $k = 1$, its readily seen that we actually have the representation

$$\tau_1(t) = \int_0^\infty \min(1, \frac{t}{s}) \min(1, s) \frac{ds}{s}.$$

Consequently, we obtain the desired representation as follows

$$\tau_1(t) = \int_0^\infty \min(1, \frac{t}{s}) \int_0^\infty \min(1, \frac{s}{r}) d\delta_1(r) \frac{ds}{s}$$

with $\delta_1(r)$ the δ measure at $r = 1$. Note that $\tau_0(s) = \min(1, s)$. For $k > 1$, we obtain, inductively, the representation

$$\tau_k(t) \approx \int_0^\infty \min(1, \frac{t}{s}) \tau_{k-1}(s) \frac{ds}{s}.$$

Therefore, by (5.3),

$$\tau_k(t) \approx \int_0^\infty \min(1, \frac{t}{s}) \min(1, s)(1 + |\log s|)^{k-1} \frac{ds}{s}$$

and we finally obtain the representation

$$\tau_k(t) \approx \int_0^\infty \min(1, \frac{t}{s}) \int_0^\infty \min(1, \frac{s}{r})[\left(\log^+ s \right)^{k-2} + \left(\log^+ \frac{1}{s}\right)^{k-2}] \frac{ds}{s}$$

Similar, elementary, but more lengthy and tedious, calculations, which shall be left to the interested reader, allow us to show that the growth functions $M(\theta) \approx \theta^{-\alpha}(1 - \theta)^{-\beta}, \alpha, \beta \geq 1$ are presentable. Moreover, the corresponding $\tau_{\alpha,\beta}(t) = \inf_\theta t^\theta \theta^{-\alpha}(1 - \theta)^{-\beta}$, can be represented by

$$\tau_{\alpha,\beta}(t) \approx \int_0^\infty \min(1, \frac{t}{s}) \int_0^\infty \min(1, \frac{s}{r})[\left(\log^+ r \right)^{\alpha-2} + \left(\log^+ \frac{1}{r}\right)^{\beta-2}] \frac{dr}{r} \frac{ds}{s}$$

On the other hand, if $\beta = 0$, say, then

$$\tau_\alpha(t) \leq c \int_0^\infty \min(1, \frac{t}{s}) \int_0^\infty \min(1, \frac{s}{r}) \left(1 + \log^+ r\right)^{\alpha-2} \frac{dr}{r} \frac{ds}{s}$$

and consequently (cf. (5.2)),

$$\tau_\alpha(t) \leq c \int_0^\infty \min(1, \frac{t}{r})(2 + \left|\log \frac{t}{r}\right|) \left(1 + \log^+ r\right)^{\alpha-2} \frac{dr}{r}. \tag{5.4}$$

Lemma 37 *Let $\bar{A}, \bar{B}$, be mutually closed pairs, and let T be a bilinear operator, defined on $\Delta(\bar{A}) \times \Delta(\bar{B})$ with values on a pair $\bar{C}$. Suppose further that*

$$T : \bar{A}_{\theta,1;J} \times \bar{B}_{\theta,1;J} \xrightarrow{M(\theta)} \bar{C}_{\theta,\infty;K}, 0 < \theta < 1,$$

with $M(\theta) \approx \theta^{-\alpha}$ as $\theta \to 0$, $\alpha \geq 1$. Then,

$$K(t, T(a, b), \bar{C}) \leq$$

$$\begin{cases} c \int_0^\infty K(s, a; \bar{A}) K(\frac{t}{s}, b; \bar{B}) \frac{ds}{s} & \alpha = 1 \\ \\ c \int_0^\infty \int_0^\infty K(s, a; \bar{A}) K(\frac{t}{sr}, b; \bar{B}) \left(1 + \log^+ r\right)^{\alpha - 2} \frac{ds}{s} \frac{dr}{r} & \alpha \geq 1 \end{cases}$$

Proof. Let $\tau_\alpha(t) = \inf_\theta t^\theta \theta^{-\alpha}$, then by (5.4) and Theorem 34 we have, if $\alpha > 1$,

$$K(t, T(a, b), \bar{C}) \leq c \int_0^\infty \int_0^\infty K(s, a; \bar{A}) K(\frac{t}{sr}, b; \bar{B}) \left(1 + \log^+ r\right)^{\alpha - 2} \frac{dr}{r} \frac{ds}{s}$$

On the other hand, when $\alpha = 1$ we have

$$K(t, T(a, b), \bar{C}) \leq c \int_0^\infty \int_0^\infty K(s, a; \bar{A}) K(\frac{t}{s}, b; \bar{B}) \frac{ds}{s}$$

as desired. $\square$

We shall now state and prove a bilinear version of Yano's extrapolation theorem.

Theorem 38 *Suppose that T is a bilinear operator such that*

$$T : \bar{A}_{\theta,1;J} \times \bar{B}_{\theta,1;J} \overset{M(\theta)}{\to} \bar{C}_{\theta,\infty;K}, 0 < \theta < 1,$$

with $M(\theta) \approx \theta^{-\alpha}$ as $\theta \to 0$, $\alpha \geq 1$. We also assume that the pairs $\bar{A}, \bar{B}$, and $\bar{C}$, are ordered and mutually closed, and moreover that $\bar{A}, \bar{B}$, are regular. Then,

$$T : \bar{A}_{(\alpha)} \times \bar{B}_{(\alpha)} \to C_0$$

Proof. We consider first the case $\alpha = 1$. We compute, using Lemma 37,

$$\|T(a, b)\|_{C_0} = K(1, T(a, b); \bar{C}) \leq c \int_0^\infty K(s, a; \bar{A}) K(\frac{1}{s}, b; \bar{B}) \frac{ds}{s}$$

We split the last integral in two parts: from 0 to 1, and from 1 to ∞, and we estimate each of them. Then, we obtain,

$$\int_0^1 K(s,a;\bar{A})K(\frac{1}{s},b;\bar{B})\frac{ds}{s} \leq \|b\|_{B_0} \int_0^1 K(s,a;\bar{A})\frac{ds}{s}$$

$$\leq \|b\|_{\bar{B}_{(1)}}\|a\|_{\bar{A}_{(1)}}$$

and,

$$\int_1^\infty K(s,a;\bar{A})K(\frac{1}{s},b;\bar{B})\frac{ds}{s} \leq \|a\|_{A_0} \int_1^\infty K(\frac{1}{s},b;\bar{B})\frac{ds}{s}$$

$$\leq \|a\|_{\bar{A}_{(1)}}\|b\|_{\bar{B}_{(1)}}$$

as required.

Consider now the case $\alpha > 1$. To proceed we need to integrate by parts. Observe that, for an ordered pair, the associated K functional of any element is constant for $t \geq 1$, therefore the derivatives of all the K functionals involved will be zero for $t > 1$. Moreover, by the regularity of the pairs $\bar{A}$, and $\bar{B}$, we have that the K functional, for any element in the corresponding spaces, satisfies $\lim_{t\to 0} K(t) = 0$. Also, to simplify the notation, we shall let $k(s) = \frac{d}{ds}(K(s))$. By Lemma 37, we have

$$\|T(a,b)\|_{C_0} = K(1,T(a,b);\bar{C})$$

$$\leq \int_0^\infty \int_0^\infty K(s,a;\bar{A})K(\frac{1}{sr},b;\bar{B})(\log r)^{\alpha-2}\frac{dr}{r}\frac{ds}{s}$$

$$\leq \int_0^\infty K(\frac{1}{s},a;\bar{A})\int_1^\infty (\log^+ r)^{\alpha-2}K(\frac{s}{r},b;\bar{B})\frac{dr}{r}\frac{ds}{s} = I.$$

Changing variables

$$I = \int_0^\infty K(\frac{1}{s},a;\bar{A})\int_0^s (\log \frac{s}{u})^{\alpha-2}K(u,b;\bar{B})\frac{du}{u}\frac{ds}{s}.$$

Integrating by parts, and using the fact that $k(u,b;\bar{B}) = 0$, $u > 1$, we have the estimate

$$I \leq c\int_0^\infty K(\frac{1}{s},a;\bar{A})\int_0^1 (\log \frac{s}{u})^{\alpha-1}k(u,b;\bar{B})du\frac{ds}{s}$$

now, interchange the order of integration and repeat the same steps to deal with $K(\frac{1}{s}, a; \bar{A})$, we get

$$I \leq c \int_0^1 k(u, b; \bar{B}) \int_0^1 k(s, a; \bar{A})(\log^+ \frac{1}{us})^\alpha ds.$$

The estimate $(\log^+ \frac{1}{us})^\alpha \leq (1 + \log \frac{1}{s})^\alpha (1 + \log \frac{1}{u})^\alpha$, implies

$$I \leq c \int_0^1 k(u, b; \bar{B})(1 + \log \frac{1}{u})^\alpha du \int_0^1 k(s, a; \bar{A})(1 + \log \frac{1}{s})^\alpha ds$$

which again integrating by parts can be seen to imply

$$I \leq c\|a\|_{(\alpha)}\|b\|_{(\alpha)}$$

as we wished to show. $\square$

As an example we write down a bilinear version of Yano's theorem,

Example 39 *(cf. [58]) Suppose that $T : L^p(0,1) \times L^P(0,1) \to L^1(0,1)$, with norm $M_p \leq c/(p-1)^\alpha, \alpha \geq 1$, as $p \to 1$, then $T : LLog^\alpha L \times LLog^\alpha L \to L^1$.*

As we have seen before a convenient way to summarize the information of the interpolation /extrapolation process is via K/J inequalities. For a bilinear operator T a K/J inequality, with function τ, is an inequality of the form:

$$K(t, T(a, b); \bar{C}) \leq \|a\|_0\|b\|_0 \tau(\frac{\|a\|_0}{\|a\|_1}, \frac{\|b\|_0}{\|b\|_1}, t)$$

Let us state, without proof, an equivalent condition to those in Theorem 33 in terms of K/J inequalities

Theorem 40 *(cf. [58]) Any of the conditions of Theorem 33 is equivalent to the K/J inequality*

$$K(t, T(a, b); \bar{C}) \leq \tau(\frac{t}{su})J(s, a, \bar{A})J(s, b, \bar{B})$$

Using these methods one can , of course, deal with more general decay rates than powers. We shall not, however, pursue the matter any further here (cf. [58]). It is also interesting to observe that these methods allow us to prove a somewhat more general version of the usual bilinear interpolation theorem of Lions-Peetre-Zafran.

Example 41 *(cf. [58]) Let $\bar{A}, \bar{B}, \bar{C}$, be pairs of spaces with $\bar{A}$ mutually closed, and let ρ, σ, γ, be quasi-concave functions. Suppose that there exists a constant $c > 0$, such that $\forall\ s, u > 0$, we have $\rho(s)\gamma(u) \leq c\sigma(su)$. Suppose further that $\frac{1}{r} = \frac{1}{p} + \frac{1}{q} - 1, 1 \leq p, q, r \leq \infty$. Then, if T is a bilinear operator such that $T : \bar{A} \times \bar{B} \to \bar{C}$, we have $T : \bar{A}_{\rho,p;K} \times \bar{B}_{\gamma,q;J} \to \bar{C}_{\sigma,r;K}$. The point is that the assumptions on T imply, as it is readily verified, the K/J estimate*

$$J(us, T(a, b); \bar{C}) \leq c \min\{1, \frac{t}{us}\} J(u, a; \bar{A}) J(s, b; \bar{B}).$$

5.2 Ideals of Operators

Let H be a Hilbert space, let S_∞ be space of bounded operators from H to H. The Schatten ideals of operators S_p are defined as follows. A compact operator $T \in S_\infty$ is in the Schatten ideal S_p, if the sequence of eigenvalues $\{s_n(T)\}$, arranged in decreasing order, of $(T^*T)^{1/2}$, belong to l^p. In this case we write

$$\|T\|_{S_p} = \| \{s_n(T)\} \|_{l^p}$$

Similarly, we say that $T \in S_{p,\infty}$ if $\{s_n(T)\} \in l(p, \infty)$.

In the literature, a class of operator ideals, the so called Macaev ideals, has been singled out and studied, as suitable end point ideals for the scale of Schatten ideals S_p. The Macaev ideals are defined as follows

$$S_w = \{T \in S_\infty : \|T\|_{S_w} = \sum_{n=1}^{\infty} \frac{s_n(T)}{n} < \infty\}$$

and

$$S_M = \{T \in S_\infty : \|T\|_{S_M} = \sup_{m \geq 1} \frac{\sum_{n=1}^{m} s_n(T)}{\sum_{n=1}^{m} \frac{1}{n}} < \infty\}$$

Now, it is well known that (cf. [65])

$$K(t, T; S_1, S_\infty) \approx \sum_{n=1}^{[t]} s_n(T) \qquad (5.5)$$

where $[t]$ is the integer part of t. We want to work with the ordered pair (S_∞, S_1) and therefore we write

$$K(t, T; S_\infty, S_1) = tK(\frac{1}{t}, T; S_1, S_\infty) = t \sum_{n=1}^{[\frac{1}{t}]} s_n(T)$$

and therefore

$$\int_0^1 K(t, T; S_\infty, S_1)\frac{dt}{t} = \int_0^1 \sum_{n=1}^{[\frac{1}{t}]} s_n(T) dt$$

$$= \int_1^\infty \sum_{n=1}^{[u]} s_n(T)\frac{du}{u^2}$$

$$= \sum_{m=1}^\infty \frac{1}{m^2} \sum_{n=1}^m s_n(T)$$

$$= \sum_{n=1}^\infty s_n(T) \sum_{m=n}^\infty \frac{1}{m^2}$$

$$\equiv \sum_{n=1}^\infty \frac{s_n(T)}{n}$$

In other words,

$$S_w = (S_\infty, S_1)_{0,1;K} \qquad (5.6)$$

Now, after the usual renormalizations, we have,

$$(S_\infty, S_1)_{\frac{1}{p}, p:K} = (S_1, S_\infty)_{\frac{1}{p'}, p;K} = S_p$$

Thus, combining (5.6) with the theory of Chapter 2, we have (with $p = \theta^{-1}$)

$$\sum_p p(S_\infty, S_1)_{\frac{1}{p}, p:K} = \sum_\theta \frac{(S_\infty, S_1)_{\theta, p(\theta):K}}{\theta} = S_w$$

Moreover, if $\rho(t) = t(1 + \log\frac{1}{t})$, $t \leq 1$, then

$$\sup_{t \leq 1} \frac{K(t, T; S_\infty, S_1)}{\rho(t)} \equiv \sup_{t \leq 1} \frac{t \sum_{n=1}^{[\frac{1}{t}]} s_n(T)}{\rho(t)} \equiv \|T\|_{S_M}$$

which combined with extrapolation theory gives

$$\Delta_p(p-1)(S_\infty, S_1)_{\frac{1}{p}, p:K} = \Delta_\theta(1-\theta)(S_\infty, S_1)_{\theta, p(\theta):K} = S_M$$

Thus, the S_w, and S_M ideals play the role in the theory of ideals of the $LLogL$ and e^L classes. The following result is a consequence of the previous discussion.

Theorem 42 *(cf. [58]) Let $T : S_p \to S_p$, then*
  (i) if $\|T\|_{S_p \to S_p} \leq p$ as $p \to \infty$, then

$$T : S_w \to S_\infty$$

(ii) if $\|T\|_{S_p \to S_p} \leq (p-1)^{-1}$ as $p \to 1$, then

$$T : S_1 \to S_M$$

Of course we can consider other extrapolation classes of ideals of operators to accommodate different rates of decay. This could be used, for example, to derive an extension of Cwikel's theorem in the next section.

It is also interesting to remark that the theory of ideals presented here can be carried out in the setting of ideals of compact operators on arbitrary Banach spaces.

5.3 Limiting case of Cwikel's estimate

We now turn to our main application. We shall extrapolate from a theorem due to Cwikel and derive from it the end point estimate obtained by Constantin [22]. Our approach suggests further extensions of Cwikel's theorem.

We shall be dealing with $L(p, q)$ spaces. In what follows we use the notation $\|.\|_{p,q}$ to denote the usual Lorentz norms, and $\|.\|_{p,q}^*$ to

denote the usual quasi-norms defined using the distribution function
or the rearrangement instead of the double star.

We consider operators acting on $L^2(T^d)$, defined by

$$T_{f,g}(h) = g((\frac{-i}{2\pi})\nabla)(fh)$$

This means that $T_{f,g}$ is formally defined on $(\varphi, \hat{\psi})$ by the inner
product of $\bar{f}\varphi$ and $(g\hat{\psi})$. The main result of [28] states that for
$2 < p < \infty$, we have

Theorem 43 *The following estimates hold*

$$\|T_{f,g}\|_{S_{p,\infty}} \leq c_p\|f\|_p\|g\|_{p,\infty}^*,$$

with $c_p = \frac{2}{p}(\frac{4}{\frac{p}{2}-1})^{1-\frac{2}{p}}(1 + \frac{2}{p-2})^{\frac{1}{p}}.$

In order to formulate this result in term of interpolation spaces
we need to be very specific about which norms we are actually con-
sidering. Using (5.5) and the reiteration formula of Holmstedt we
see that the K functional for the pair (S_2, S_∞) can be computed by
(cf. [65]):

$$K(N, T; S_2, S_\infty) \approx \left\{\sum_{k=1}^{N^2} s_k(T)^2\right\}^{\frac{1}{2}} \tag{5.7}$$

Thus, we can write $S_{p,\infty} = (S_2, S_\infty)_{\theta,\infty;K}$, that is

$$\|H\|_{S_{p,\infty}} = \sup_N \{N^{-\theta}\left\{\sum_{n=1}^{N^2} s_k(T)^2\right\}^{\frac{1}{2}}\}, \quad 1 - \frac{2}{p} = \theta.$$

Let us also write

$$L^{p,\infty} = (L^{2,\infty}, L^\infty)_{\theta,\infty;K}, \quad \theta = 1 - \frac{2}{p}$$

Then, since

$$\|f\|_{(L^{2,\infty}, L^\infty)_{\theta,\infty;K}} \geq c\|f\|_{p,\infty}^*$$

where $\theta = 1 - \frac{2}{p}$, we see that the estimate of Theorem 43 implies

$$\|T_{f,g}\|_{(S_2, S_\infty)_{\theta,\infty;K}} \leq c_\theta\|f\|_{(L^2, L^\infty)_{\theta,\infty;K}}\|g\|_{(L^{2,\infty}, L^\infty)_{\theta,\infty;K}}$$

where $c_\theta = (1 - \theta)^{-1}(4(1 - \theta)\theta^{-1})^\theta(1 + (1 - \theta)\theta^{-1})^{(1-\theta)/2}$.

Summing up, when $\theta \to 0$, we have arrived to the estimate

$$\|T_{f,g}\|_{(S_2,S_\infty)_{\theta,\infty;K}} \leq \theta^{-\frac{1}{2}}\|f\|_{(L^2,L^\infty)_{\theta,\infty;K}}\|g\|_{(L^{2,\infty},L^\infty)_{\theta,\infty;K}} \qquad (5.8)$$

Taking the infimum we may now obtain the K/J inequalities for the bilinear operator $T_{f,g}$. The corresponding function τ is defined by

$$\tau(s,u,t) = \inf_{\theta \in (0,1)} \{\theta^{-1/2}(\frac{t}{su})^\theta\} \leq 1 + e^{1/2}\sqrt{2}\log_+^{1/2}(\frac{t}{su}) \qquad (5.9)$$

Therefore (5.8), (5.9), and Theorem 40, give

$$K(N,T_{f,g},S_2,S_\infty) \leq c\|f\|_2\|g\|_{L_{2,\infty}}\{1 + \log_+^{1/2}(\frac{N\|f\|_\infty\|g\|_\infty}{\|f\|_2\|g\|_{L_{2,\infty}}})\}$$

and recalling the estimate for the K functional for the pair (S_2, S_∞) given in (5.7) and making the change of variable $N \to N^{1/2}$ we arrive to

$$\{\sum_{i=1}^{N} s_k(T_{f,g})^2\}^{1/2} \leq c\|f\|_2\|g\|_{2,\infty}(1 + \log_+^{1/2}(N^{1/2}\frac{\|f\|_\infty\|g\|_\infty}{\|f\|_2\|g\|_{2,\infty}}))\}$$

$$(5.10)$$

This is the statement of Constantin's theorem in [22] except for the fact that he states his theorem for vector valued functions. However, since Theorem 43 extends mutatis mutandis to the vector valued setting we obtain, likewise, the full strength of the result of [22].

The reader of these notes will have no problem in deriving the end point version of Cwikel's theorem that can be obtained from (5.10) (or equivalently (5.8)) and can be framed in terms of variants of the Macaev ideals discussed in the previous section.

5.4 Notes and Further Results

The relationship between extrapolation and approximation theory is considered in [57], [18] and [58]. For a pair $\bar{A}$, we let

$$F(t,a;\bar{A}) = \begin{cases} \infty & \text{if } \|a\|_{A_1} > t \\ \|a\|_{A_0} & \text{if } \|a\|_{A_1} \leq t \end{cases}$$

and define

$$\bar{A}_{\theta,q;F} = \{x : x = \int_0^\infty u(s)\frac{ds}{s},\ F(s, u(s); \bar{A}) < \infty; \|x\|_{\bar{A}_{\theta,q;F}} < \infty\}$$

with

$$\|x\|_{\bar{A}_{\theta,q;F}} = \inf_{x=\int_0^\infty u(s)\frac{ds}{s}} \left\{ \left\{ \int_0^\infty (F(t, a; \bar{A})t^{-\theta})^q \frac{dt}{t} \right\}^{\frac{1}{q}} \right\}$$

These spaces coincide with the usual approximation/interpolation spaces for the usual range of the parameters. At the end points of the scales we have modify the definitions somewhat. Working with a "discrete" definition we let, for an ordered pair $\bar{A}$,

$$\|x\|_{\bar{A}_{0,q;F}} = \inf\{\left\{\sum_{n=1}^\infty 2^n (F(2^{2^n}, u_n))^q\right\}^{\frac{1}{q}} : x = \sum_{n=1}^\infty u_n,\ F(2^{2^n}, u_n) < \infty\}$$

and

$$\bar{A}_{0,q;F} = \{x : \|x\|_{\bar{A}_{0,q;F}} < \infty\}$$

Similarly, for the E functional defined by

$$E(t, x; \bar{A}) = \inf\{\|x - x_1\|_{A_0} : \|x_1\|_{A_1} \le t\}$$

we let

$$\bar{A}_{0,q;E} = \{x : \left\{\sum_{n=1}^\infty n^{-1}(E(n, x; \bar{A}))^q\right\}^{\frac{1}{q}} < \infty\}$$

Then (cf. [18], [58]),

$$\bar{A}_{0,q;E} = \bar{A}_{0,q;F}$$

We cannot go into the details here and we refer to [58] and the references quoted therein. For example, in the setting of operator ideals these results imply a description of the elements of Macaev ideals in terms of the rate at which they can be approximated by operators of finite dimensional rank. For example,

$$S_w = \{T : T = \sum_{n=1}^\infty T_n, \text{with } rank(T_n) \le 2^{2^n}, \sum_{n=1}^\infty 2^n \|T_n\|_{S_\infty} < \infty\}.$$

However, the point we want to make is that the formal analysis works in general to describe extrapolation spaces in terms of degree of approximation by suitable subspaces. Thus, for example, using appropriate descriptions of Besov spaces as approximation spaces (cf. [8]) one can characterize extrapolation spaces for Besov scales in a similar fashion (for example operators of finite dimensional rank could be replaced here by entire functions of exponential type, etc). We think that these descriptions could be useful in the study of compression of wavelet decompositions and its applications to image compression. In this setting the subspaces from which we approximate are finite dimensional spaces generated by wavelets. For references to work in this area we refer to [38], [35] and [59]. There is forthcoming related work by Houdre (cf. [51]).

Chapter 6

Extrapolation, Reiteration, and Applications

Although, as we have stressed in these notes, extrapolation spaces can be incorporated in an extended theory of interpolation spaces, they enjoy special properties which distinguish them from the classical spaces introduced by Lions and Peetre. A central issue in interpolation theory is the idea of reiteration. In this chapter we give several general reiteration theorems, for extrapolation spaces obtained by the $\sum$ method, which are motivated by applications to classical analysis. In particular, we consider an application to the study of the of the maximal operator of partial sums of Fourier series and show that certain limiting inequalities can be obtained through a combination of reiteration and extrapolation techniques. We also consider applications to the theory of logarithmic Sobolev inequalities. We show how to derive the so called higher order logarithmic Sobolev inequalities from lower order estimates, and we indicate new estimates for scales of spaces *closer* to L^2 than the $L^2(LogL)^n$ scale.

6.1 Reiteration

In this section we consider some simple, and useful, reiteration properties which are needed to treat our applications.

Let us recall a characteristic property of the $\sum$ functor:

$$\sum_\theta \varphi(\theta)\mathcal{F}_\theta(\bar{A}) = \sum_{\theta < \theta_0} \varphi(\theta)\mathcal{F}_\theta(\bar{A})$$

where $\bar{A}$ is an ordered pair, $\mathcal{F}_\theta$ exact functor, and $\frac{1}{\varphi(\theta)} \in L^\infty$. Moreover, if the pair $\bar{A}$ is mutually closed then we have,

$$\sum_\theta \varphi(\theta)\mathcal{F}_\theta(\bar{A}) = \sum_\theta \varphi(\theta)\bar{A}_{\theta,1;J} = \bar{A}_{\rho,1;J},$$

with $\rho(t) = \sup \frac{t^\theta}{\varphi(\theta)}$. In particular, if $\varphi(\theta) = \theta^{-\alpha}$, with $\alpha \geq 0$, we have $\rho(t) = \rho_\alpha(t)$ (cf. (2.17)), and

$$\sum_\theta (\theta^{-\alpha}\bar{A}_{\theta,1;J}) = \sum_{\theta<\theta_0} (\theta^{-\alpha}\bar{A}_{\theta,1;J}) = \bar{A}_{\rho_\alpha,1;J} = \bar{A}_{\rho_{\alpha-1},1;K}$$

Now, by the reiteration theorem, for the real method of interpolation, we get

$$\sum_{\theta<\theta_0} (\theta^{-\alpha}\bar{A}_{\theta,1;J}) = \sum_\theta (\theta^{-\alpha}(A_0, \bar{A}_{\theta_0,1;J}))_{\theta_0\theta,1;J}$$

$$= (A_0, \bar{A}_{\theta_0,1;J})_{\rho_\alpha,1;J}$$

Thus, combining the results above, we get

$$\bar{A}_{\rho_\alpha,1;J} = \bar{A}_{\rho_{\alpha-1},1;K} = (A_0, \bar{A}_{\theta_0,1;J})_{\rho_\alpha,1;J}$$

$$= (A_0, \bar{A}_{\theta_0,1;J})_{\rho_{\alpha-1},1;K}$$

More generally, for any $0 < \theta_0 < 1,\, 0 < q \leq \infty$,

$$\bar{A}_{\rho_\alpha,1;J} = \bar{A}_{\rho_{\alpha-1},1;K} = (A_0, \bar{A}_{\theta_0,q})_{\rho_\alpha,1;J}$$

$$= (A_0, \bar{A}_{\theta_0,q})_{\rho_{\alpha-1},1;K}$$

This type of *constancy* was discovered in [43], in the context of the K method: $\forall q \in (0,\infty],\, \forall \theta \in (0,1)$,

$$(A_0, A_1)_{(1);K} = (A_0, \bar{A}_{\theta,q})_{(1);K} \qquad (6.1)$$

The novelty here is given by the fact that we are allowed to vary both parameters: (θ, q) at the same time. Recall that in the classical reiteration theorem (cf. [8]) we have that, if $0 < p, q, r \leq \infty$, then $\forall$ $\theta, \mu \in (0, 1)$,

$$(A_0, \bar{A}_{\theta, q})_{\mu;p} = (A_0, \bar{A}_{\theta, r})_{\mu;p}.$$

The corresponding limiting property for the J method has been known for a long time, namely $\forall q \in (0, \infty], \forall \theta \in (0, 1)$,

$$(A_0, A_1)_{0,1;J} = (A_0, \bar{A}_{\theta, q})_{0,1;J} = A_0^\circ$$

Results of this type are very useful in the actual computation of extrapolation spaces as we shall show in this chapter.

It will be convenient in what follows to use the following notation. Let $\bar{A}$ be an ordered pair, $0 < p \leq \infty$, then we let

$$\bar{A}_{0,p;K} = \left\{ x : x \in A_0, \left\{ \int_0^1 [K(t, x; \bar{A})]^p \frac{dt}{t} \right\}^{1/p} < \infty \right\}$$

thus $\bar{A}_{0,p;K} = \bar{A}_{(1);K}$, as defined in (2.20).

Example 44 *Consider the following proof (cf. [43]) of a well known interpolation theorem by Zygmund: If T is a linear operator defined on $L^1(0, 1)$ which is of weak type $(1,1)$ and of type (p, p) for some $p > 1$, then T maps $LLogL$ into L^1. In fact, by hypothesis*

$$T : (L^1, L^p)_{0,1;K} \to (wL^1, wL^p)_{0,1;K}$$

Now, by the constancy property

$$(L^1, L^p)_{0,1;K} = (L^1, L^\infty)_{0,1;K} = LLogL$$

while it is easy to see that

$$(wL^1, wL^p)_{0,1;K} \subset L^1$$

In a similar vein one can extend to general interpolation scales the following interpolation theorem by O'Neil [84].

Example 45 *If T is a sublinear operator of weak types $(1, p)$, (q, r), with $0 < p < r < \infty$, $1 \leq q < \infty$, then for $0 < \theta < 1$, $T : L(LogL)^\theta \to L^{p, 1/\theta}$.*

Proof.   We apply the $(.,.)_{0,1;K}$ method as in Example 44. Then, since

$$(L^{p,\infty}, L^{r,\infty})_{0,1;K} = (L^{p,\infty}, L^{\infty})_{0,1;K} \subset L^{p,1}$$

we get

$$T : LLogL \to L^{p,1}$$

Now, this estimate can be interpolated once again with

$$T : L^1 \to L^{p,\infty}$$

and we obtain

$$T : (L^1, L(LogL))_{\theta,1} \to (L^{p,\infty}, L^{p,1})_{\theta,1}$$

But, since (cf. [7])

$$L^1, L(LogL))_{\theta,1} = L(LogL)^\theta,$$

and (Hölder's inequality),

$$(L^{p,\infty}, L^{p,1})_{\theta,1} \subset L^{p,1/\theta}$$

we are done. □

We recall other related end point reiteration theorems from [43]. Let $\bar{A}$ be an ordered pair of (quasi-) Banach spaces, then it is shown in [43] that

$$K(t, f; A_0, (A_0, A_1)_{0,1;K}) \approx t \int_{e^{-\frac{1}{t}}}^{1} K(u, f; \bar{A}) \frac{du}{u} \qquad (6.2)$$

and

$$K(t, f; (A_0, A_1)_{0,1;K}, A_1) \approx \int_0^1 K(\min\{\varphi^{-1}(t), u\}, f; \bar{A}) \frac{du}{u} \qquad (6.3)$$

where $\varphi(t) = t \, \log(e/t)$.

We now prove a more general version of (6.3) which we require for our application to logarithmic Sobolev inequalities. We refer the reader to [43] for a proof of (6.2).

Theorem 46 *Let $\bar{A}$ be an ordered Banach pair, $0 < p \leq \infty$, then*

$$K(t, f; \bar{A}_{0,p;K}, A_1) \approx \left\{ \int_0^1 \left(K(\min\{\varphi^{-1}(t), u\}, f; \bar{A}) \right)^p \frac{du}{u} \right\}^{\frac{1}{p}}$$

where $\varphi(t) = t[\log \frac{e}{t}]$.

Proof. The proof is patterned after the corresponding one in [43]. Let $f \in A_0$, $t > 0$, and let

$$L(f)(t) = \left\{ \int_0^1 \left(K(\min\{\varphi^{-1}(t), u\}, f; \bar{A}) \right)^p \frac{du}{u} \right\}^{\frac{1}{p}}.$$

Suppose that $f = f_0 + f_1$, $f_i \in A_i$, $i = 0, 1$, then

$$L(f)(t) \leq L(f_0)(t) + L(f_1)(t).$$

Now, since the K functional is increasing we trivially have

$$L(f_0)(t)^p \leq \int_0^1 [K(s, f_0; \bar{A})]^p \frac{ds}{s} = \|f\|_{\bar{A}_{0,p;K}}^p$$

To estimate the second term we use the trivial estimate $K(s, f_1; \bar{A}) \leq s\|f_1\|_{A_1}$, to obtain,

$$L(f_1)(t) \leq \|f_1\|_{A_1} \left\{ \int_0^1 \left(\min\{\varphi^{-1}(t), u\} \right)^p \frac{du}{u} \right\}^{\frac{1}{p}}$$

Splitting the last integral appropriately yields

$$\left\{ \int_0^1 \left(\min\{\varphi^{-1}(t), u\} \right)^p \frac{du}{u} \right\}^{\frac{1}{p}} = p^{-\frac{1}{p}} \varphi^{-1}(t)^{\frac{1}{p}} + \varphi^{-1}(t) \log\left(\frac{e}{\varphi^{-1}(t)} \right)^{\frac{1}{p}}$$

Now, since

$$\varphi^{-1}(t) [\log\left(\frac{e}{\varphi^{-1}(t)} \right)]^{1/p} = (\varphi^{-1}(t))^{1-1/p} [\varphi^{-1}(t) \log\left(\frac{e}{\varphi^{-1}(t)} \right)]^{1/p}$$

$$= (\varphi^{-1}(t))^{1-1/p} t^{1/p} \leq t$$

(where have used in the last step that $\varphi^{-1}(t) \leq t$), we get

$$L(f_1)(t) \leq ct\|f_1\|_{A_1}$$

Consequently, we have obtained

$$L(f)(t) \leq cK(t, f; \bar{A}_{0,p;K}, A_1)$$

Conversely, let $f = f_0(t) + f_1(t)$ be an optimal decomposition of f for the pair $\bar{A}$, i.e. such that $K(t, f; \bar{A}) \approx \|f_0(t)\|_{A_0} + t\|f_1(t)\|_{A_1}$. We claim that the decomposition

$$f = f_0(\varphi^{-1}(t)) + f_1(\varphi^{-1}(t))$$

is optimal for the pair $(\bar{A}_{0,p;K}, A_1)$. In fact,

$$\left\| f_0(\varphi^{-1}(t)) \right\|_{\bar{A}_{0,p;K}} = I + II$$

where

$$I = \left\{ \int_0^{\varphi^{-1}(t)} K(s, f_0(\varphi^{-1}(t)); \bar{A})^p \frac{ds}{s} \right\}^{\frac{1}{p}}$$

$$II = \left\{ \int_{\varphi^{-1}(t)}^1 K(s, f_0(\varphi^{-1}(t)); \bar{A})^p \frac{ds}{s} \right\}^{\frac{1}{p}}$$

Writing $f_0(\varphi^{-1}(t)) = f - f_1(\varphi^{-1}(t))$, and using the triangle inequality, we get

$$I \leq L(f)(t) + \left\{ \int_0^{\varphi^{-1}(t)} K(s, f_1(\varphi^{-1}(t)); \bar{A})^p \frac{ds}{s} \right\}^{\frac{1}{p}}$$

$$\leq L(f)(t) + \left\| f_1(\varphi^{-1}(t)) \right\|_{A_1} \left\{ \int_0^{\varphi^{-1}(t)} s^p \frac{ds}{s} \right\}^{\frac{1}{p}}$$

$$\leq L(f)(t) + p^{-1/p} K(\varphi^{-1}(t), f; \bar{A}).$$

Now, since $K(s, f)/s$ decreases, we have

$$K(\varphi^{-1}(t), f; \bar{A}) \leq c \left\{ \int_0^{\varphi^{-1}(t)} \left(K(s, f; \bar{A}) \right)^p \frac{ds}{s} \right\}^{\frac{1}{p}} \leq cL(f)(t)$$

and therefore

$$I \leq cL(f)(t)$$

A similar argument also yields

$$II \leq cL(f)(t).$$

Combining these two estimates we obtain,

$$\left\| f_0(\varphi^{-1}(t)) \right\|_{\bar{A}_{0,p;K}} \le c\, L(f)(t)$$

On the other hand, since $t \in (0,1)$,

$$t\| f_1(\varphi^{-1}(t))\|_{A_1} \le t\, K(\varphi^{-1}(t), f; \bar{A}) \le K(\varphi^{-1}(t), f; \bar{A}) \le c L(f)(t).$$

We have thus obtained

$$\left\| f_0(\varphi^{-1}(t)) \right\|_{\bar{A}_{0,p;K}} + t\| f_1(\varphi^{-1}(t))\|_{A_1} \le c\, L(f)(t)$$

which implies

$$K(t, f, \bar{A}_{0,p;K}, A_1) \le c L(f)(t)$$

and the result follows. □

The following application of the previous theorem will be useful later.

Theorem 47 *Let $\bar{A}$ be an ordered pair, then*

$$((A_0, A_1)_{0,p;K}, A_1)_{0,p;K} = \{ f : \int_0^1 \left(K(s, f; \bar{A}) \right)^p \left(\log \frac{1}{s} \right)^p \frac{ds}{s} < \infty \}$$

In particular,

$$((A_0, A_1)_{0,1;K}, A_1)_{0,1;K} = (A_0, A_1)_{(2);K}$$

Proof. To simplify the notation we let $p = 1$. By (6.3) (the proof for $p > 1$ is the same, we just have to use Theorem 46 instead)

$$K(t, f; A_0, A_1)_{0,1;K}, A_1) \approx K(\varphi^{-1}(t), f; \bar{A}) \ln \frac{1}{\varphi^{-1}(t)} + \int_0^{\varphi^{-1}(t)} K(s, f; \bar{A}) \frac{ds}{s}$$

Thus, integrating with respect to $\frac{dt}{t}$ on $(0,1)$ we find

$$\| f \|_{((A_0, A_1)_{0,1;K}, A_1)_{0,1;K}} \approx$$

$$\int_0^1 K(\varphi^{-1}(t), f; \bar{A}) \ln \frac{1}{\varphi^{-1}(t)} \frac{dt}{t} + \int_0^1 \int_0^{\varphi^{-1}(t)} K(s, f; \bar{A}) \frac{ds\, dt}{s\, t}$$

To evaluate the first integral on the right hand side we make the change of variable $u = \varphi^{-1}(t)$, then we get

$$\int_0^1 K(\varphi^{-1}(t), f; \bar{A}) \ln \frac{1}{\varphi^{-1}(t)} \frac{dt}{t} \approx \int_0^1 K(s, f; \bar{A}) \left(1 + \log \frac{1}{s}\right) \frac{ds}{s}$$

The remaining integral is computed interchanging the order of integration, and we see that

$$\int_0^1 \int_0^{\varphi^{-1}(t)} K(s, f; \bar{A}) \frac{ds\, dt}{s\, t} = \int_0^1 K(s, f; \bar{A}) \left(\log \frac{1}{s} + \log \log \frac{e}{s}\right) \frac{ds}{s}$$

and the result follows. $\square$

Another computation of extrapolation spaces, that can be obtained by these methods, is the following extension of Theorem 6. (The reader should compare the methods of proof of these two results).

Theorem 48 (*cf. [72]*)

$$((A_0, A_1)_{(1);K}, (A_0, A_1)_{(2);K})_{(1);K} =$$

$$\{f \in (A_0, A_1)_{(1);K} : \int_0^1 K(s, f; \bar{A})[\log^+(\log \frac{1}{s})] \frac{ds}{s} < \infty\}$$

Proof.   The proof is by "ping-pong" iteration. In fact successively applying Theorem 47, (6.2), and (6.3) (applied to $B_0 = (A_0, A_1)_{(1);K}$; $B_1 = (B_0, A_1)_{(1);K}$) we get

$$K(t, f; (A_0, A_1)_{(1);K}, (A_0, A_1)_{(2);K}) \approx$$

$$K(t, f; (A_0, A_1)_{(1);K}, ((A_0, A_1)_{(1);K}, A_1)_{(1);K})$$

$$\approx t \int_{e^{-\frac{1}{t}}}^1 K(u, f; (A_0, A_1)_{(1);K}, A_1) \frac{du}{u}$$

$$\approx t \int_{e^{-\frac{1}{t}}}^1 \int_0^1 K(\min\{\varphi^{-1}(u), s\}, f; \bar{A}) \frac{ds}{s} \frac{du}{u}$$

Therefore,

$$\|f\|_{((A_0, A_1)_{(1);K}, (A_0, A_1)_{(2);K})_{(1);K}} \approx$$

$$\approx \int_0^1 \int_{e^{-\frac{1}{t}}}^1 \int_0^1 K(\min\{\varphi^{-1}(u), s\}, f; \bar{A}) \frac{ds}{s} \frac{du}{u} \, dt$$

which in turn is equivalent to the sum of two integrals

$$= (I) + (II), \text{ say.}$$

where,

$$(I) = \int_0^1 K(s, f; \bar{A}) \int_{\varphi(s)}^1 \int_0^{\frac{1}{\log \frac{1}{u}}} dt \, \frac{du}{u} \frac{ds}{s}$$

$$(II) = \int_0^1 \int_{e-1/t}^1 K(\varphi^{-1}(u), f; \bar{A}) \log \frac{1}{\varphi^{-1}(u)} \frac{du}{u} \, dt$$

Now,

$$(II) \leq \int_0^1 \left[\frac{K(\varphi^{-1}(u), f; \bar{A})}{\varphi^{-1}(u)} \right] \left[\varphi^{-1}(u) \log \frac{e}{\varphi^{-1}(u)} \right] (\log \frac{1}{u})^{-1} \frac{du}{u}$$

$$= \int_0^1 \left[\frac{K(\varphi^{-1}(u), f; \bar{A})}{\varphi^{-1}(u)} \right] (\log 1/u)^{-1} du$$

$$= \int_0^1 \left[\frac{K(u, f; \bar{A})}{u} \right] (\log \frac{1}{\varphi(u)})^{-1} \varphi'(u) du$$

$$\leq \|f\|_{(A_0, A_1)_{(1); K}}$$

On the other hand,

$$(I) = \int_0^1 K(s, f; \bar{A}) \int_{\varphi(s)}^1 \frac{1}{\log \frac{1}{u}} \frac{du}{u} \frac{ds}{s}$$

$$\leq \int_0^1 K(s, f; \bar{A}) \log^+ \left(\log \frac{1}{\varphi(s)} \right) ds \approx$$

$$\int_0^1 K(s, f, \bar{A}) \log^+ \left(\log^+ \left(\log \frac{1}{s} \right) \right) ds + \int_0^1 K(s, f, \bar{A}) \log^+ \left(\log^+ \frac{1}{s} \right) ds$$

and the desired result follows. □

Example 49 *Consider the pair $\bar{A} = (L^1(T), L^\infty(T))$. Then, $\bar{A}_{(1); K} = LLogL(T)$, $\bar{A}_{(2); K} = L(LogL)^2(T)$, and, by Theorem 48, $(\bar{A}_{(1); K}, \bar{A}_{(2); K})_{(1); K} = LLogLLogLogL(T)$. Combining this calculation with $L(LogL(T))^{1+\theta} = [\bar{A}_{(1); K}, \bar{A}_{(2); K}]_\theta$ (see (7.15)) and (2.19) we obtain Theorem 6.*

6.2   Estimates for the Maximal Operator Of Partial Sums Of Fourier Series

We give a summary of the relevant estimates for the maximal operator on partial sums of Fourier Series. The fundamental results in this area are due to Carleson [17] and Hunt [52].

Theorem 50 *Let $S(f) = \sup_n |S_n(f)|$, where $S_n(f)$ denotes the nth partial sum of the Fourier series of f, $1 < p < \infty$, then for every f of the form, $f = g\chi_F$, with $2^{-1} < g \le 1$, we have*

$$\sup_{t>0} t^{\frac{1}{p}}(Sf)^*(t) \le c_p \|f\|_p$$

where $c_p = O(\frac{1}{p-1})$.

Combining Theorem 50, the fact that $L^{p,\infty}(T) \subset L^1(T)$, with norm $O(\frac{1}{p-1})$ as $p \to 1$, and the extrapolation theorem of Yano, allows Hunt to conclude (cf. [52])

Theorem 51 $S : L(Log)^2(T) \to L^1(T)$.

These results were improved by Carleson and Sjölin (cf. [94]), and many other authors. I refer to the papers [57], and [72], for a more detailed bibliography on this. In particular, relevant to the methods developed in these notes, chapter 5 of [57] deals with extrapolation of weak type estimates in this context.

In [95] the following result is proved

Theorem 52 *If $f \in L(LogL)^{1+\theta}(T)$ then $Sf \in L(LogL)^{\theta-1}(T)$, $0 < \theta \le 1$.*

A complement of this result was obtained in [97]

Theorem 53 *If $f \in LLogL(LogLogL)(T)$ then $Sf \in L(LogL)^{-1}(T)$.*

A perusal of the constants in the proof of Theorem 52 in [95] shows that

$$\int_T Sf(t)(1 + \log^+ Sf(t)))^{\theta-1}dt \le C\theta^{-1}\int_T |f(t)|(1 + \log^+|f(t)|)^{1+\theta}dt$$

$$(6.4)$$

where C is a constant independent of θ.

6.3   Extrapolation Methods

In this section, which is based on [72], we give a proof of Theorem 53 by extrapolation. In fact, we shall show that Theorem 53 can be extrapolated from (6.4).

Let us first observe that

$$L(LogL)^{\theta-1}(T) \subset L(LogL)^{-1}(T)$$

In fact, we have

$$\int_T |f(t)|[1 + \log^+ |f(t)|]^{-1} dt \le \int_T |f(t)|[1 + \log^+ |f(t)|]^{\theta-1} dt \quad (6.5)$$

We also need to quantify the relationship between different ways of measuring the size of a function in the $L(LogL)^\beta(T)$ spaces.

Lemma 54 *For $\theta \in (0, \frac{1}{2})$, let*

$$l_\theta(f) = \|f\|_{L(LogL)^{1+\theta}(T)} = \int_0^1 f^*(t)[1 + \log \frac{1}{t}]^{1+\theta} dt.$$

Then, we have

$$\int_0^1 |f(t)|[1 + \log^+ |f(t)|]^{1+\theta} dt \le c\Gamma_\theta(f)$$

where c is a constant independent of θ, and

$$\Gamma_\theta(f) = \begin{cases} l_\theta(f) & \text{if } l_\theta(f) \le 1 \\ \frac{[l_\theta(f)]^2}{\theta} & \text{if } l_\theta(f) > 1 \end{cases} \quad (6.6)$$

Proof.   We may suppose that $l_\theta = l_\theta(f) < \infty$. It is readily seen that

$$f^*(t) \le l_\theta t^{-1}$$

Therefore, we obtain

$$\int_0^1 |f(t)|[1+\log^+ |f(t)|]^{\theta+1} dt = \int_{\{f^*(t) \le l_\theta t^{-1}\}} f^*(t)[1+\log^+ f^*(t)]^{1+\theta} dt$$

$$\le \int_0^1 f^*(t)[1 + \log^+ \frac{l_\theta}{t}]^{1+\theta} dt \quad (6.7)$$

We consider two cases. If $l_\theta \leq 1$, then we estimate the function $[1 + \log^+ \frac{l_\theta}{t}]^{1+\theta}$ separately in $(0, l_\theta)$ and $(l_\theta, 1)$, and we see that the right hand side of (6.7) is bounded by $4\,l_\theta$. On the other hand, if $l_\theta > 1$, then

$$\int_0^1 f^*(t)[1 + \log^+ \frac{l_\theta}{t}]^{1+\theta} dt \leq (1 + \log\, l_\theta)^{1+\theta} l_\theta.$$

Now, since $\log\, l_\theta \leq \theta^{-1} l_\theta^\theta$, and $l_\theta > 1$, $\theta \in (0, \frac{1}{2})$, we have

$$(1 + \log\, l_\theta)^{1+\theta} l_\theta \leq (1 + \frac{l_\theta}{\theta})^{1+\theta} l_\theta \leq (2\,l_\theta)^{1+\theta} \theta^{-1} \theta^{-\theta} \leq c(l_\theta)^2 \theta^{-1} \theta^{-\theta}.$$

The desired result follows since $\theta \in (0, 1/2)$, and $\theta^{-\theta}$ is bounded as $\theta \to 0$. $\square$

Using (6.4), (6.5) and Lemma 54, we have,

$$S : L(LogL)^{1+\theta}(T) \to L(LogL)^{-1}(T).$$

with

$$\int_T |Sf(t)|[1 + \log^+ |Sf(t)|]^{-1} dt \leq c\theta^{-1} \Gamma_\theta(f) \qquad (6.8)$$

as $\theta \to 0$.

· Note that the functional

$$f \to l(f) = \int_T |f(t)|[1 + \log^+ |f(t)|]^{-1} dt \qquad (6.9)$$

is subaditive on positive functions. Let us also note that $Sf \geq 0$, and that, moreover, S is a subaditive operator. If a norm estimate were available instead of (6.4) we could extrapolate directly. However, in the situation at hand we need an extra argument to overcame the nonlinearity of the estimate. Let $f \in \sum_{\theta \in (0,1)} \{\theta^{-1} L(LogL)^{1+\theta}(T)\} = \sum_{\theta \in (0,\frac{1}{2})} \{\theta^{-1} L(LogL)^{1+\theta}(T)\}$. Consider a nearly optimal decomposition $f = \sum_{\theta < \frac{1}{2}} f_\theta$ such that

$$\|f\|_{\sum_\theta \{\theta^{-1} L(LogL)^{1+\theta}(T)\}} \approx \sum_{\theta < \frac{1}{2}} \theta^{-1} \|f_\theta\|_{L(LogL)^{1+\theta}(T)}$$

We further decompose f as follows. Let

$$f = f_0 + f_1, \quad \text{with } f_0 = \sum_{\theta \in E} f_\theta, f_1 = \sum_{\theta \in F} f_\theta$$

where $E = \{\theta \in (0, \frac{1}{2}) : \|f_\theta\|_{L(LogL)^{1+\theta}(T)} \leq 1\}$, $F = \{\theta \in (0, \frac{1}{2}) : \|f_\theta\|_{L(LogL)^{1+\theta}(T)} > 1\}$.

Applying the functional l defined in (6.9) to Sf, and taking into account the sublinearity of the functionals, we get

$$l(Sf) \leq l(Sf_0) + l(Sf_1)$$

We now estimate each of these terms separately. By subadditivity,

$$l(Sf_0) \leq \sum_{\theta \in E} l(Sf_\theta)$$

$$\leq c \sum_{\theta \in E} \theta^{-1} \Gamma_\theta(f_\theta) \text{ (by (6.8))}$$

$$\leq c \sum_{\theta \in E} \theta^{-1} l_\theta(f_\theta) \text{ (by the definitions of } E \text{ and } \Gamma)$$

$$\leq c \sum_{\theta \in (0, \frac{1}{2})} \theta^{-1} \|f_\theta\|_{L(LogL)^{\theta+1}(T)} \approx \|f\|_{\sum_\theta \{\theta^{-1} L(LogL)^{1+\theta}(T)\}}$$

Similarly, since the functional $f \to [l(f)]^{\frac{1}{2}}$ is also subadditive on positive functions, we obtain

$$[l(Sf_1)]^{\frac{1}{2}} \leq c \sum_{\theta \in F} [l(Sf_\theta)]^{\frac{1}{2}}$$

$$\leq c \sum_{\theta \in F} \theta^{-\frac{1}{2}} [\Gamma(f_\theta)]^{\frac{1}{2}} \text{ (by (6.8))}$$

$$\leq c \sum_{\theta \in F} \theta^{-\frac{1}{2}} \theta^{-\frac{1}{2}} \|f_\theta\|_{L(LogL)^{1+\theta}(T)} \text{ (by(6.8))}$$

$$\leq c \sum_{\theta \in (0, \frac{1}{2})} \theta^{-1} \|f_\theta\|_{L(LogL)^{1+\theta}(T)} \approx \|f\|_{\sum_\theta \{\theta^{-1} L(LogL)^{1+\theta}(T)\}}$$

Consequently,

$$l(Sf) \leq c\{\|f\|_{\sum_\theta \{\theta^{-1} L(LogL)^{1+\theta}(T)\}} + \left(\|f\|_{\sum_\theta \{\theta^{-1} L(LogL)^{1+\theta}(T)\}}\right)^2\}$$

as we wished to show. $\square$

6.4   More On Reiteration

In this section we collect some auxiliary results needed for our applications to higher order logarithmic Sobolev inequalities below. We consider some calculations with the $(.,.)_{0,q;K}$ methods. In fact, one could prove similar results for the more general $(.,.)_{\rho_\alpha,q;K}$ methods, but we shall not pursue the most general results here.

A typical calculation with the $(.,.)_{0,p;K}$ functor is given by the following

Example 55 *Let Ω be a probability space, then*

$$(L^p(\Omega), L^\infty(\Omega))_{0,p;K} = L^p LogL(\Omega)$$

Proof.   *This follows readily from the well known fact (cf. [8]) that*

$$(K(t, f; L^p, L^\infty))^p \approx \int_0^{t^p} [f^*(s)]^p ds$$

(a proof can be obtained by reiteration and (2.28)). Integrating with respect to $\frac{dt}{t}$ on $(0,1)$ the result follows. $\square$

Likewise, one can also prove that

$$(L^p(\Omega), L^\infty(\Omega))_{0,p;K} = (L^p(\Omega), L^r(\Omega))_{0,p;K} \ , \ p < r. \tag{6.10}$$

More generally, we have

Lemma 56 *Let $\bar{A}$ be an ordered Banach pair, then*

$$(A_0, \bar{A}_{\theta,q})_{0,p;K} = (A_0, A_1)_{0,p;K}$$

Proof.   The proof follows mutatis mutandis the corresponding result for $p = 1$. In fact, it is enough to show that $\forall \theta \in (0,1)$ we have $(A_0, \bar{A}_{\theta,\infty;K})_{0,p;K} \subset \bar{A}_{0,p;K}$. Suppose that $f \in (A_0, \bar{A}_{\theta,\infty;K})_{0,p;K}$, then by Holmstedt's formula (cf. [8]) we have

$$K(t, f; A_0, \bar{A}_{\theta,\infty;K}) \approx t \sup_{t^{\frac{1}{\theta}} < s} s^{-\theta} K(s, f; \bar{A}) \geq K(t^{1/\theta}, f; \bar{A}).$$

Therefore,

$$\int_0^1 \left(K(t,f;A_0,\bar{A}_{\theta,\infty;K})\right)^p \frac{dt}{t} \geq \int_0^1 \left(K(t^{1/\theta},f;\bar{A})\right)^p \frac{dt}{t}$$

and the result follows. $\square$

We shall not pursue the standard extension of the results presented for $p = 1$ to the case $p > 1$. However, for our application to Logarithmic Sobolev inequalities we do require the following

Example 57 $(L^2 LogL, L^\infty)_{0,2;K} = L^2(LogL)^2$. *(All spaces in this example are based on a probability space Ω).*

Proof. *In fact using successively Example 55, Theorem 47, and the estimate of the K functional in Example 55 we obtain,*

$$((L^2 LogL, L^\infty)_{0,2;K} = ((L^2, L^\infty)_{0,2;K}, L^\infty)_{0,2;K} =$$

$$= \{f : \int_0^1 \int_0^{t^2} f^*(s)^2 \log \frac{1}{t} \frac{dt}{t} ds < \infty\}$$

$$= \{f : \int_0^1 f^*(s)^2 \int_{s^{\frac{1}{2}}}^1 \log \frac{1}{t} \frac{dt}{t} ds < \infty\}$$

$$= \{f : \int_0^1 f^*(s)^2 (\log \frac{1}{s})^2 ds < \infty\}$$

$$= L^2(LogL)^2$$

$\square$

6.5 Higher Order Logarithmic Sobolev Inequalities

We consider some aspects of the logarithmic Sobolev inequalities of Gross [46] and the higher order estimates of Feisner [41]. Our contribution here is to show that extrapolation spaces can be used to provide a simple proof of the higher order estimates from the lower order ones. The twist is that this time the extrapolation spaces are obtained by interpolation. This gives a new approach to the estimates by Feisner [41], a related method based in the theory of commutators has been given in [25]. Our method can be also applied to derive new results for logarithmic spaces that are closer to L^2 than $L^2(LogL)^n$ type spaces. We discuss this briefly in the notes at the end of the chapter. A detailed study of the relationship between extrapolation and logarithmic Sobolev inequalities shall given elsewhere (cf. [78]).

Let us introduce some notation and background. Our presentation here follows Bakry and Meyer [3]. Let μ be the standard gaussian measure on R, and let P_t be the semigroup of Hermite-Ornstein-Uhlenbeck,

$$P_t = \int_R f(e^{-t/2}x + (1 - e^{-t})^{1/2}y)\mu(dy).$$

Then, $P_t = \int_{R_+} e^{-\lambda t}dE_\lambda$, with E_λ discrete and concentrated on $\{\frac{n}{2} : n = 0, 1, ...\}$. The Riesz potentials associated with P_t are defined for $Re(v) \geq 0$, by

$$R^v = \int_{(0,\infty)} \lambda^{-v}dE_\lambda,$$

which correspond to the L^2 multipliers $\lambda^{-v}\chi_{\{\lambda>0\}}(= \lambda^{-v}\chi_{\{\lambda\geq1/2\}}$, in dE_λ), and are therefore bounded). It can be shown that, in fact

$$R^v f = \Gamma(v)^{-1} \int_0^\infty t^{v-1}P_t f \, dt.$$

In what follows all the spaces considered will based on $d\mu$, unless otherwise specified.

The following fundamental result is due to Nelson.

Theorem 58 *Let* $1 \leq p \leq q \leq 1 + e^t(p - 1)$, *then* P_t *is a bounded contraction from* L^p *to* L^q.

A weak version (due to loss of information on the constants) of the logarithmic Sobolev inequalities of Gross can be now stated as follows

Theorem 59 $R^{1/2}$ *is a bounded operator from* L^2 *to* L^2LogL.

Our purpose in this section is to use Theorem 60 and reiteration theorems for extrapolation spaces in order to prove the so called Higher Order Logarithmic Sobolev Inequalities, due to Feissner (cf. [41]).

Theorem 60 *For every $n \in N, R^{1/2}$ is a bounded map,*

$$R^{1/2} : L^2(LogL)^n \rightarrow L^2(LogL)^{n+1}$$

Proof. The case $n = 0$ is Theorem 60. To prove the case $n = 1$, we use the case $n = 0$, and the trivially verified (Minkowski's inequality) fact that $R^{1/2} : L^\infty \rightarrow L^\infty$. Applying the $(.,.)_{0,2;K}$ method we get

$$R^{1/2} : (L^2, L^\infty)_{0,2;K} \rightarrow (L^2 LogL, L^\infty)_{0,2;K}$$

The desired result now follows from the identifications

$$(L^2, L^\infty)_{0,2;K} = L^2 LogL \text{ (by (6.10))}$$

and

$$(L^2 LogL, L^\infty)_{0,2;K} = L^2(LogL)^2, \text{ by Example 57}$$

An induction argument concludes the proof for $n \geq 2$. $\square$

Remark. A simple modification of these ideas leads to a proof of Feissner's results for $p \geq 2$.

6.6 Notes and Further Results

A systematic study of the reiteration properties of extrapolation spaces is given in [48]. In particular it is shown there how the Lorentz-Zygmund spaces can be obtained via reiteration theorems for extrapolation spaces.

Recently, several authors (cf. [40]) have derived variants of the Sobolev imbedding theorem involving "double exponential integrability". We wish to indicate very briefly here how to approach such results via reiteration theorems for extrapolation spaces. Using the method outlined in the proof of Theorem 46 (cf. [43]) one can prove that, if $\bar{A}$ is an ordered pair, then

$$K(t, f, \bar{A}_{\rho,\infty;K}, A_1) \approx \sup_{0 < s \leq e^{1-\frac{1}{t}}} \left\{ \frac{K(s, f, A_0, A_1)}{\rho(s)} \right\}.$$

where $\rho(t) = \rho_1(t) = t(1 + \log \frac{1}{t})$. (A similar result holds for ρ_α).

Now, recall that

$$(L^1, L^\infty)_{\rho,\infty;K} = e^L$$

Thus, we have:

$$f \in (e^L, L^\infty)_{\rho,\infty;K} \iff$$

$$\sup_{t\in(0,1)} \left\{ \frac{1}{\rho(t)} \sup_{0<s\le e^{1-\frac{1}{t}}} \frac{\int_0^s f^*(u)du}{\rho(s)} \right\} < \infty$$

$$= \sup_{s\in(0,1)} \left\{ \frac{1}{\rho(s)} \int_0^s f^*(u)du \sup_{t>\frac{1}{1+\log\frac{1}{s}}} \left\{ \frac{1}{t(1+\log\frac{1}{t})} \right\} \right\} < \infty$$

$$= \sup_{s\in(0,1)} \left\{ \frac{1}{\rho(s)} \int_0^s f^*(u)du(1+\log\frac{1}{s}) \frac{1}{1+\log(1+\log\frac{1}{s})} \right\} < \infty$$

$$\sup_{s\in(0,1)} \left\{ f^{**}(s) \frac{1}{1+\log(1+\log\frac{1}{s})} \right\} < \infty$$

It follows that

$$(e^L, L^\infty)_{\rho,\infty;K} = e^{e^L}.$$

Thus, interpolating in this fashion, known extreme Sobolev embedding theorems (cf. Chapter 4, Theorem 32, and the discussion following it), yields results of the type sought after in [40]. We hope to return to this point elsewhere.

We indicate now an improvement to the logarithmic Sobolev estimates that follows through the use of the methods developed in this chapter. We try to get closer and closer to the space L^2, therefore we interpolate the estimates of Gross and Feissner using an extrapolation method,

$$R^{1/2} : (L^2, L^2(LogL))_{0,2;K} \to (L^2 LogL, L^2(LogL)^2)_{0,2;K}$$

The corresponding interpolation spaces can be identified through the use of techniques similar to those developed in the text. For example, the required extension of the reiteration formula (6.2) is given by

$$K(t, f, A_0, (A_0, A_1)_{0,p;K}) \approx t \left\{ \int_{e^{1/tp}}^1 (K(s, f; \bar{A}))^p \frac{ds}{s} \right\}^{1/p} \qquad (6.11)$$

An extension of Theorem 48, and Theorem 47, now yields,

$$R^{1/2} : L^2 Log(LogL) \to L^2(LogL)(Log(LogL)).$$

We can continue by iteration,

$$R^{1/2} : L^2 \underbrace{Log(Log(...LogL)}_{n \text{ times}} \to L^2(LogL)(\underbrace{Log(Log(...LogL))}_{n \text{ times}}$$

For this and other related results we refer to [78].

These results should be compared to those available by regular interpolation, namely

$$R^{1/2} : L^2(LogL)^\theta \to L^2(LogL)^{1+\theta}.$$

Chapter 7

Estimates For Commutators In Real Interpolation

In this chapter we study interpolation theorems, for certain possibly nonlinear operators, where cancellation plays an important role. The results are formulated in terms of commutators between a given bounded operator and certain operators Ω which are associated with interpolation scales in a natural way. The typical example we have in mind can be described as follows. Suppose that K is a Calderón-Zygmund operator, and for $b \in BMO$, let M_b be the operator defined by $M_b f = bf$, then (cf. [?]) the operator

$$[K, M_b] = KM_b - M_b K$$

satisfies

$$\|[K, M_b]f\|_{L^p} \leq c\|f\|_{L^p} \ , 1 < p < \infty.$$

Note that each of the individual terms in the commutator is unbounded in L^p while the cancellation effect of the operation $[.,.]$ produces a bounded operator. This result is important in the theory of compensated of Murat and Tartar, as developed by Coifman, P.L. Lions, Meyer and Semmes in [19]. For example, it can be used to prove Theorem 24.

Rochberg and Weiss [89], and Jawerth, Rochberg and Weiss [60] have shown that associated with the classical methods of interpolation there are certain operators, Ω, generally unbounded and non-linear, such that if T is a bounded operator in the scale, then the commutator $[\Omega, T]$, is also bounded in the scale. Again this boundedness is due to cancellation since each of the individual terms of the commutator is unbounded. For example, for $(L^p(w_0), L^p(w_1))$ a possible choice for Ω is $\Omega f = f \log(\frac{w_0}{w_1})$, this fact combined with the relationship between the space BMO and the theory of A_p weights gives the commutator theorem of Coifman-Rochberg-Weiss alluded above (cf. Example 69 below).

The connection with compensated compactness goes further. Recently Greco and Iwaniec (cf. [44]) and the author (cf. [74], [75]) have established, through the use of commutator techniques, new sharp estimates for Jacobians. Iwaniec and Sbordone [55] have obtained new estimates for perturbations of Hodge type decompositions for vector fields, with interesting applications to the study of the regularity of solutions to variational problems, in particular the regularity of A—harmonic maps. Further results in this direction were recently given in [69]. For further development of the relationship between compensated compactness and interpolation theory we refer to [44], [74], [76], and the references therein.

The theory has also been applied to the study of H^p spaces in the work of Kalton (cf. [61], [62]); logarithmic Sobolev inequalities in [25]; ideals of operators in [61].

More recently, Rochberg [88], and the author (cf. [77]), have obtained higher order commutator estimates for the complex and real methods of interpolation. In the higher order theory the degree of unboundedness of the individual terms is greater as is the degree of the non-linearity. On the other hand, the higher amount of cancellation present allows one to obtain control on the norm of the higher order commutators. It is hoped that these new results will also find interesting applications.

In this chapter we shall indicate some of the main lines of these developments. We also develop new connections of the abstract theory of commutators to extrapolation theory, and the functional calculus associated with a positive operator in a Banach space. We believe that our results in this direction could have applications in the theory of parabolic equations. We also formulate, through the use of extrapolation spaces, some new end point results.

7.1 Some Operators Associated to Optimal Decompositions

In this section we show how to construct operators in interpolation scales through the use of optimal decompositions.

It will be convenient for our purposes here to introduce some special notation. For $\theta \in R$, $0 < q \leq \infty$, and for measurable functions on the half line, let

$$\phi_{\theta,q}(f) = \left\{ \int_0^\infty \left| f(t)t^{-\theta} \right|^q \frac{dt}{t} \right\}^{\frac{1}{q}}$$

Thus, the interpolation spaces $\bar{A}_{\theta,q;K}$, $0 < \theta < 1$, $0 < q \leq \infty$, are defined by

$$\bar{A}_{\theta,q;K} = \{ f \in \Sigma(\bar{A}) : \|f\|_{\bar{A}_{\theta,q;K}} = \phi_{\theta,q}\left(K(., f, \bar{A})\right) < \infty \} \quad (7.1)$$

Let us say that a decomposition $f = f_0(t) + f_1(t)$ is almost optimal, for the K method, if

$$\|f_0(t)\|_{A_0} + t\|f_1(t)\|_{A_1} \leq cK(t, a; \bar{A}) \quad (7.2)$$

where c is a constant fixed before hand, say $c = 2$. We then write $D_K(t)f = D_K(t; \bar{A})f = f_0(t)$. For $n \in N$, the operators $\Omega_{\bar{A},n;K} = \Omega_{n;K}$ associated with this decomposition are defined by

$$\Omega_{n;K} f = \frac{1}{(n-1)!} \left(\int_0^1 (\log t)^{n-1} D_K(t) f \frac{dt}{t} - \int_1^\infty (\log t)^{n-1} (I - D_K(t)) f \frac{dt}{t} \right) \quad (7.3)$$

Similarly, one defines the corresponding operators Ω associated with the J and E methods. For a Banach pair $\bar{A}$, let us say that $u(t)$ is an almost optimal decomposition of f for the J method, if

$$f = \int_0^\infty u(s) \frac{ds}{s}, \quad \|f\|_{\bar{A}_{\theta,q;J}} \approx \phi_{\theta,q}\left(J(., u(.), \bar{A})\right) \quad (7.4)$$

In which case we write $D_J(t)f = u(t)$, and the corresponding $\Omega_{\bar{A},n;J} = \Omega_{n;J}$ operators are defined by

$$\Omega_{n;J} f = \frac{1}{n!} \int_0^\infty D_J(t) f (\log t)^n \frac{dt}{t} \quad (7.5)$$

For the E method we have a similar definition. Recall that

$$E(t, f; \bar{A}) = \inf_{\|f_1\|_{A_1} \leq t} \{\|f_0\|_{A_0} : f = f_0 + f_1\}$$

The corresponding interpolation spaces $\bar{A}_{\theta,q;E}$, $0 < \theta < \infty$, $0 < q \leq \infty$, are defined using the quasi-norms

$$\|f\|_{\bar{A}_{\theta,q,E}} = \phi_{-\theta,q}\left(E(., f, \bar{A})\right) \tag{7.6}$$

We say that $f_0(t)$ is an almost optimal decomposition, for the E method, and we write $D_E(t)f = D_E(t; \bar{A})f = f_0(t)$, if

$$E(t, f; \bar{A}) \approx \|D_E(t)f\|_{A_0} \tag{7.7}$$

Then, for $n \in N$, we let $\Omega_{\bar{A},n;E} = \Omega_{n;E}$ be defined by

$$\Omega_{n;E}f = \frac{1}{(n-1)!} \left(\int_1^\infty (\log t)^{n-1} D_E(t)f \frac{dt}{t} \right.$$
$$\left. - \int_0^1 (\log t)^{n-1}(I - D_E(t))f \frac{dt}{t} \right) \tag{7.8}$$

Whenever no confusion arises we shall drop the subindex that indicates which method of interpolation we are using in the definition of the $\Omega's$.

We give examples of computations of these operators in different settings.

Example 61 *We consider in detail the pair* (L^1, L^∞). *For the E method (cf. [57]) we get an optimal decomposition writing*

$$f = f_0(t) + f_1(t), \, f_0(t) = (f - t) \chi_{\{|f|>t\}}, \, f_1(t) = t\chi_{\{|f|>t\}} + f\chi_{\{|f|\leq t\}}$$

Then, we have

$$\Omega_{n,E}f = \frac{1}{n!}f\left(\log|f|\right)^n - \frac{1}{(n-1)!}\int_0^{|f|}(\log t)^{n-1}dt,$$

It is however more convenient to consider the almost optimal decomposition (cf. [60])

$$f = f\chi_{\{|f|<t\}} + f\chi_{\{|f|\leq t\}}$$

to obtain as another possible choice for $\Omega_{n,E}$:

$$\Omega_{n,E}f = \frac{1}{n!}f\left(\log|f|\right)^n$$

Example 62 *If we use the K method then the same considerations of [60] for $n = 1$, lead to the general formula*

$$\Omega_{n,K}f(x) = \frac{1}{n!}f(x)\left(\ln(r_f(x))\right)^n$$

where the "rank function" $r_f(x)$ is defined by

$$r_f(x) = |\{y : |f(y| > |f(x)| \ \ or \ \ |f(y| = |f(x)| \,, y \leq x\}|$$

Example 63 *The pair (L^{p_0}, L^{p_1}). Let $\bar{A}$ be a pair, Holmstedt's proof of the reiteration theorem (cf. [8]) implies that for $\bar{B} = (\bar{A}_{\theta_0,q_0;K}, \bar{A}_{\theta_1,q_1;K})$ we have*

$$\Omega_{n,\bar{B},K} = |\theta_0 - \theta_1|^n \, \Omega_{n,\bar{A},K} \tag{7.9}$$

and a similar result holds for the E method (cf. [60], [25]). This remark shows, in particular, that we can compute Ω_K for (L^{p_0}, L^{p_1}) using Example 62.

Example 64 *The pair $(L^p(w_0), L^p(w_1))$. Let $X = X(0, \infty)$ be a Banach lattice. A positive function $w(t)$ defined on $(0, \infty)$ shall be called a "weight". We can use weights to construct "weighted" versions of X. Namely, let X_w denote the space of measurable functions on $(0, \infty)$ such that $fw \in X$, with the norm given by*

$$\|f\|_{X_w} = \|fw\|_X$$

Let w_i be weights, $i = 0, 1$, then, it is well known, and easy to see, that

$$K(t, f, X_{w_0}, X_{w_1}) \approx \|f \min\{w_0, tw_1\}\|_X$$

In fact, a nearly optimal decomposition of f is given by

$$f = f\chi_{\{w_0 \leq tw_1\}} + f\chi_{\{w_0 > tw_1\}}$$

Thus an Ω_K associated with the pairs (X_{w_0}, X_{w_1}) is given by

$$\Omega f = f \log \frac{w_0}{w_1}$$

For more details see [60]. More generally for these pairs we have,

$$\Omega_{n;K} f = cf \left(\log \frac{w_0}{w_1} \right)^n \tag{7.10}$$

Similar results hold for the complex method, in fact Rochberg [88] shows that, for the pair $(L^p(w_0), L^p(w_1))$, one can choose

$$\Omega_{n,C} f = \frac{1}{n!} f \left(\log(\frac{w_0}{w_1}) \right)^n$$

In the case of L^p spaces we can also define the weighted spaces $L^p(wdx)$. Observe that since $L^p(w) = L^p(w^{1/p}dx)$, we have

$$\Omega_{(L^p(w_0 dx), L^p(w_1 dx))} = \frac{1}{p} \Omega_{(L^p(w_0), L^p(w_1))}$$

as long as $p < \infty$.

Example 65 *Through the use of the Fourier transform we can reduce the computation of Ω for the pair of Sobolev spaces (H^2, H^{-2}) to the computation of Ω for the pair $(L^2((1+|x|^2)), L^2((1+|x|^2)^{-1}))$. Using the previous calculations and taking inverse Fourier transform we see that we can choose*

$$\Omega f = c \log(I + \Delta) f$$

where $\log(I + \Delta)$ is defined using the functional calculus. For more details see [60], and for the definition of concave functions of a positive operator on a Banach space see (7.44) below.

Example 66 *Through the use of the fundamental lemma, operators Ω associated with the J method can be computed from the corresponding ones for the K method (cf. [24])*

$$\Omega_{n,K} = -\Omega_{n,J} \tag{7.11}$$

The following result from [60] is the main abstract result concerning the boundedness of commutators on real interpolation scales (for the corresponding results for the complex method see [89] and [88]).

Theorem 67 *(cf. [60]) Let $\bar{A}, \bar{B}$, be Banach pairs, T be a bounded linear operator $T : \bar{A} \to \bar{B}$, and, moreover, if F denotes any of the real methods of interpolation described above, let $[T, \Omega_{1,F}] = T\Omega_{F(\bar{A}),1} - \Omega_{F(\bar{B}),1}T$. Then, there exists a constant $c = c(F)$, such that*

$$\|[T, \Omega_{1,F}]f\|_{F(\bar{B})} \le c\|f\|_{F(\bar{A})}. \tag{7.12}$$

Theorem 67 has been recently extended to higher order commutators by the author (cf. [77]).

Theorem 68 *Let $\bar{A}$ and $\bar{B}$ be a Banach pairs, $T : \bar{A} \to \bar{B}$ be a bounded operator. For $n = 0, 1, 2....$ define*

$$C_n f = \begin{cases} T & n = 0 \\ [T, \Omega_{1,K}]f & n = 1 \\ ... & ... \\ [T, \Omega_{n,K}]f + \Omega_{1,K}C_{n-1,K}f + ... + \Omega_{n-1,K}C_{1,K}f & n \ge 2 \end{cases}$$

If F denotes either the E or J methods, we define

$$C_n f = \begin{cases} T & n = 0 \\ [T, \Omega_{1,F}]f & n = 1 \\ ... & ... \\ [T, \Omega_{n,F}]f - \Omega_{1,F}C_{n-1,F}f - ... - \Omega_{n-1,F}C_{1,F}f & n \ge 2 \end{cases}$$

Then, for $0 < \theta < 1, 1 \le q \le \infty$, there exist absolute constants $c = c(\theta, q, n, T) > 0$, such that, for all the methods of interpolation above, we have

$$\|C_n f\|_{(B_0, B_1)_{\theta,q}} \le c\|f\|_{(A_0, A_1)_{\theta,q}}.$$

Let us illustrate these results with applications to singular integrals.

Example 69 *Let $b \in BMO$, and assume that b has BMO norm sufficiently small. Then, e^b and e^{-b} are A_p weights and therefore if K is a Calderón -Zygmund operator, we have, for $1 < p < \infty$,*

$$K : (L^p(e^b dx), L^p(e^{-b}dx)) \to (L^p(e^b dx), L^p(e^{-b}dx))$$

Since by Example 64 we can choose $\Omega_{(L^p(e^b dx), L^p(e^{-b} dx)); K} f = M_b f = bf$, *we get*

$$[K, M_b] : L^p = (L^p(e^b), L^p(e^{-b}))_{\frac{1}{2}, p; K} \rightarrow (L^p(e^b), L^p(e^{-b}))_{\frac{1}{2}, p; K} = L^p$$

$$\|[K, M_b] f\|_{L^p} \leq c \|f\|_{L^p}$$

An homogeneity argument (i.e. replacing b by cb) shows that the assumption on the BMO norm of b is not a restriction on b.

Example 70 *The higher order estimates that correspond to the previous example are given by*

$$[...[[K, M_b], M_b]......, M_b] : L^p \rightarrow L^p.$$

(cf. [88]).

7.2   Method of Proof

This section is devoted to the proof of Theorem 67. We shall also briefly indicate the strategy for the proof of Theorem 68, although we must refer the reader to [77] for the complete details.

We introduce further useful notation in order to formulate a result from [80] for commutators of order 1. Let P and $\tilde{P}$ be defined on measurable functions on $(0, \infty)$ by

$$P(f)(t) = \int_0^t f(s) \frac{ds}{s}, \quad \tilde{P}(f)(t) = t \int_t^\infty f(s) \frac{ds}{s^2} \tag{7.13}$$

and let

$$S(f)(t) = P(f)(t) + \tilde{P}(f)(t)$$

In this context Hardy's inequality can be stated as

$$\phi_{\theta, q}(Sf) \leq c_{\theta, q} \phi_{\theta, q}(f), 0 < \theta < 1, 1 \leq q \leq \infty \tag{7.14}$$

The following formula can be easily established by an elementary computation

$$\Omega_{n, K} f + \frac{1}{n!} (\log t)^n f \tag{7.15}$$

$$= \frac{1}{(n-1)!}\left(\int_0^t (\log s)^{n-1} D_K(s) f \frac{ds}{s} - \int_t^\infty (\log s)^{n-1}(I - D_K(s)) f \frac{ds}{s}\right)$$

In particular, for $n = 1$, we obtain

$$\Omega_{1,K} f + (\log t) f = \int_0^t D_K(s) f \frac{ds}{s} - \int_t^\infty (I - D_K(s)) f \frac{ds}{s} \quad (7.16)$$

Thus, for each $t > 0$, (7.16) provides us with a decomposition of $\Omega_{1,K} f + (\log t) f$. Let us now check that, in fact, this is a "good" decomposition, *modulo* a bounded operator. In fact, we have

$$\left\| \int_0^t D_K(s) f \frac{ds}{s} \right\|_{A_0} \leq \int_0^t \|D_K(s) f\|_{A_0} \frac{ds}{s}$$

$$\leq \int_0^t K(s, f; \bar{A}) \frac{ds}{s} \quad \text{(by definition of } D_K(s) f\text{)}$$

Similarly, we see that

$$t \left\| \int_t^\infty (I - D_K(s)) f \frac{ds}{s} \right\|_{A_1} \leq t \int_t^\infty \frac{K(s, f; \bar{A})}{s} \frac{ds}{s}$$

Consequently, we have proved that $\forall t > 0$,

$$K(t, \Omega_{1,K} f + (\log t) f; \overline{A}) \leq c S\left(K(., f; \bar{A})\right)(t) \quad (7.17)$$

where c is an absolute constant independent of t and f.

Therefore, by (7.14), we get, for $0 < \theta < 1, 1 \leq q \leq \infty$,

$$\phi_{\theta,q}(K(t, \Omega_{1,K} f + (\log t) f; \overline{A})) \leq c_{\theta,q} \|f\|_{\overline{A}_{\theta,q;K}} \quad (7.18)$$

Theorem 67, for the K method, can be now proved as follows. Let $T : \bar{A} \to \bar{B}$ be a given bounded operator, let $f \in \bar{A}_{\theta,q;K}$, and let $\Omega = \Omega_{1,K}$; we need to control

$$\|[T, \Omega] f\|_{\bar{B}_{\theta,q;K}} = \phi_{\theta,q}(K(t, [T, \Omega] f; \bar{B}))$$

To use (7.17) to the advantage we write

$$K(t, [T, \Omega] f; \bar{B}) = K(t, T(\Omega f + f \log t) - (\Omega T f + T f \log t); \bar{B})$$

$$\leq K(t, T(\Omega_1 f + f \log t); \bar{B}) + K(t, \Omega_1 T f + T f \log t; \bar{B})$$

$$\leq \|T\| \, K(t, \Omega_1 f + f \log t; \bar{A}) + K(t, \Omega_1 Tf + Tf \log t; \bar{B})$$

Consequently,

$$\phi_{\theta,q}(K(t, [T, \Omega_{1,K}]f; \bar{B}))$$

$$\leq c\phi_{\theta,q}(K(t, \Omega_1 f + f \log t; \bar{A})) + \phi_{\theta,q}(K(t, \Omega_1 Tf + Tf \log t; \bar{B}))$$

Now, we can apply (7.18) to estimate the first term,

$$\phi_{\theta,q}(K(t, \Omega_1 f + f \log t; \bar{A})) \leq c_{\theta,q} \|f\|_{\bar{A}_{\theta,q;K}}$$

Likewise, to estimate the second term we successively use (7.17), the boundedness of T, and Hardy's inequality (7.14) to obtain

$$\phi_{\theta,q}(K(t, \Omega_1 Tf + Tf \log t; \bar{B})) \leq c\phi_{\theta,q}\left(S\left(K(., Tf; \bar{B})\right)(t)\right)$$

$$\leq c\phi_{\theta,q}\left(S\left(K(., f; \bar{A})\right)(t)\right)$$

$$\leq c\|f\|_{\bar{A}_{\theta,q;K}}$$

and we are done.

In view of Example 66 the result just obtained for the K method implies the corresponding one for the J method. To complete the proof of Theorem 67 it thus remains to consider the E method. We record the following analog of (7.15)

$$\Omega_{n,E}f - \frac{1}{n!}(\log t)^n f \tag{7.19}$$

$$= \frac{1}{(n-1)!}\left(\int_t^\infty D_E(s)f(\log s)^{n-1}\frac{ds}{s} - \int_0^t (I - D_E(s))f(\log s)^{n-1}\frac{ds}{s}\right)$$

Again we consider the case $n = 1$, and we observe that the previous formula gives us a decomposition of $\Omega_{1,E}f - \log t \, f$ that allows us to compute the E functional of this element. In fact, note that

$$\left\|\int_0^t (I - D_E(s))f\frac{ds}{s}\right\|_{A_1} \leq \int_0^t \|(I - D_E(s))f\|_{A_1}\frac{ds}{s}$$

$$\leq \int_0^t s\frac{ds}{s} = t.$$

Therefore,

$$E(t, \Omega_{1,E}f - \log tf; \bar{A}) = \inf_{\|a_1\|_{A_1}\leq t} \|\Omega_{1,E}f - \log tf - a_1\|_{A_0}$$

$$\leq \left\| \Omega_{1,E} f - \log t f - \int_0^t (I - D_E(s)) f \frac{ds}{s} \right\|_{A_0}$$

$$= \left\| \int_t^\infty D_E(s) f \frac{ds}{s} \right\|_{A_0}$$

$$\leq \int_t^\infty \| D_E(s) f \|_{A_0} \frac{ds}{s}$$

$$\leq c \int_t^\infty E(s, f; \bar{A}) \frac{ds}{s} \tag{7.20}$$

where the last inequality follows by the definition of $D_E(s)f$.

We use this estimate to take advantage of the cancellation as follows. We need to control $\phi_{\theta,q}(E(t, [T, \Omega_{1,E}]; \bar{B}))$. For this purpose let us write

$$E(t, [T, \Omega_{1,E}]f; \bar{B}) = E(t, T\Omega_{1,E}f - \log tTf + \log tTf - \Omega_{1,E}Tf; \bar{B})$$

$$\leq E(t, T\Omega_{1,E}f - \log tTf; \bar{B}) + E(t, \log tTf - \Omega_{1,E}Tf; \bar{B})$$

Then, by (7.20), and a variant of Hardy's inequality, we get

$$\| [T, \Omega_{1,E}]f \|_{B_{\theta,q;E}} \leq c \| f \|_{\bar{A}_{\theta,q;E}}$$

concluding the proof of Theorem 67.

The proof of the Theorem 68 follows a similar strategy, although the details are somewhat more complicated. Let us briefly indicate the main ingredients of the method. First we introduce iterates of the operators $P, \tilde{P}, S$, as follows

$$P^{(n)}(f) = \underbrace{P(.....P(P(f)))...)}_{n \text{ times}}, \quad \tilde{P}^{(n)}(f) = \underbrace{\tilde{P}(.....\tilde{P}(\tilde{P}(f)))...)}_{n \text{ times}}$$

$$S_n(f) = P^{(n)}(f) + \tilde{P}^{(n)}(f) \tag{7.21}$$

By computation, we recognize the familiar operators studied in Chapter 2,

$$S_n(f) = \int_0^\infty \left| \log(\frac{t}{s}) \right|^{n-1} \min\{1, \frac{t}{s}\} f(s) \frac{ds}{s}$$

By iteration of (7.14) we see that these operators are bounded in the $\phi_{\theta,q}$ norms. Now, we select expressions that can be controlled by these operators. For example, in the case $n = 2$ we prove

$$K(t, (\log t)\Omega_{1,K}f - \Omega_{2,K}f + \frac{1}{2}(\log t)^2 f; \bar{A}) \leq cS_2(K(., f; \bar{A}))(t)$$

In fact, multiply (7.16) by $\log t$ to obtain

$$\log t\Omega_1 f + (\log t)^2 f = \int\limits_0^t \log t D_K(s)f\frac{ds}{s} - \int\limits_t^\infty \log t(I - D_K(s))f\frac{ds}{s}$$

$$\tag{7.22}$$

Now, (7.15) for $n = 2$ gives

$$\Omega_2 f + \frac{1}{2}(\log t)^2 f = \int\limits_0^t \log s D_K(s)f\frac{ds}{s} - \int\limits_t^\infty \log s(I - D_K(s))f\frac{ds}{s}$$

$$\tag{7.23}$$

and adding (7.22) and (7.23) we see that,

$$\log t\Omega_1 f + \frac{1}{2}(\log t)^2 f - \Omega_2 f$$

$$= \int_0^t \log\frac{t}{s}D_K(s)f\frac{ds}{s} + \int_t^\infty \log\frac{s}{t}(I - D_K(s))f\frac{ds}{s}$$

Thus, by the triangle inequality and the readily verified facts that

$$K(t, D_K(s)f; \bar{A}) \leq K(s, f; \bar{A}); K(t, (I - D_K(s))f; \bar{A}) \leq \frac{t}{s}K(s, f; \bar{A})$$

$$\tag{7.24}$$

we obtain,

$$K(t, \log t\Omega_1 f + \frac{1}{2}(\log t)^2 f - \Omega_2 f; \bar{A})$$

$$\leq c\int_0^\infty \min\{1, \frac{t}{s}\}\left|\log\frac{t}{s}\right|K(s, f; \bar{A})\frac{ds}{s}$$

$$= cS_2(K(., f; \bar{A}))(t)$$

as desired.

The general case follows thus a combinatorial pattern where the expressions to control are given by the computation of the powers of $\log\frac{t}{s} = \log t - \log s$. We obtain

$$K(t, \Omega_n f + p_n (\log t)^n f + \sum_{k=1}^{n-1} (\log t)^k \frac{(-1)^{n-1-k}}{k!} \Omega_{n-1-k} f; \bar{A}) \quad (7.25)$$

$$\le cS_n(K(., f; \bar{A}))(t)$$

where $p_n = \left(\sum_{k=1}^{n-1}(-1)^{n-1-k}\frac{1}{k!(n-k)!}\right) + \frac{1}{n!}$.

Now, the proof Theorem 68 follows from (7.25) by a suitable modification of the arguments used to prove Theorem 67 (cf. [77] for the details).

7.3 Computation of Ω for extrapolation spaces

The formula (7.9) is not valid for extrapolation spaces. But using the proofs of the reiteration theorems given in Chapter 6 it is not difficult to formulate appropriate substitutes. In this section we indicate in detail how to do so, and compute the corresponding operators Ω associated with certain extrapolation spaces. For a more detailed study of commutator theorems in extrapolation scales, and in particular in Orlicz spaces near L^1, and further applications we refer to [45].

Let $\bar{A}$ be an ordered pair, then the corresponding operator $\Omega = \Omega_K$ for the pair $(A_0, \bar{A}_{0,1;K})$ can be computed using a result of [43], which, for convenience, we state here as follows

$$D_K(s, A_0, \bar{A}_{0,1;K})f = D_K(e^{-\frac{1}{s}}, A_0, A_1)f = D_K(e^{-\frac{1}{s}})f \quad (7.26)$$

Thus, the corresponding operator associated with the pair $(A_0, \bar{A}_{0,1;K})$, which we shall denote $\Omega^{(-1)}$, can be computed,

$$\Omega^{(-1)}f = \int_0^1 D_K(e^{-\frac{1}{s}}, A_0, A_1)f\frac{ds}{s} = \int_0^{e^{-1}} D_K(s)f\frac{ds}{s \ln\frac{1}{s}}$$

More generally, the proof of (6.11) leads to the formula

$$D_K(s, A_0, \bar{A}_{0,p;K})f = D_K(e^{-\frac{1}{sp}}, A_0, A_1)f = D_K(e^{-\frac{1}{sp}})f$$

and therefore the operator Ω associated with the pair $(A_0, \bar{A}_{0,p;K})$, $p > 0$, is given by,

$$\Omega f = \int_0^1 D_K(e^{-\frac{1}{sp}}, A_0, A_1) f \frac{ds}{s} = \frac{1}{p} \int_0^{e^{-1}} D_K(s) f \frac{ds}{s \left(\ln \frac{1}{s} \right)^{1/p}}.$$

Example 71 *In particular, for the pair* (L^1, L^∞), *over a probability measure space, the operator* $\Omega^{(-1)} = \Omega$, *associated with the pair* $(L^1, L(LogL)) = (L^1, (L^1, L^\infty)_{0,1;K})$ *is given by*

$$\Omega^{(-1)} f(x) = f(x) \ln \ln r_f(x)$$

where $r_f(x)$ *is defined in Example 62 above.*

This process can be continued, thus we form

$$A_{-n} = \begin{cases} A_1 & n = 0 \\ (A_0, A_1)_{0,1;K} & n = 1 \\ (A_0, (A_0, A_1)_{0,1;K})_{0,1;K} & n = 2 \\ \quad \cdots & \cdots \\ (A_0, A_{-(n-1)})_{0,1;K} & n \end{cases}$$

The spaces A_{-n} can be computed using the nth iterate of the (7.26) which leads to the following extension of the reiteration formula (6.2):

$$K(t, f; A_0, A_{-n}) \simeq t \int_{e^{-\frac{1}{t}}}^1 \int_{e^{-\frac{1}{u_n}}}^1 \cdots \int_{e^{-\frac{1}{u_2}}}^1 K(u_1, f; A_0, A_1) \frac{du_1}{u_1} du_2 .. du_n$$

$$(7.27)$$

For example, if $n = 2$, then

$$K(t, f, A_0, A_{-2}) \approx t \int_{e^{-e^{1/t}}}^1 K(u_1, f; A_0, A_1) \frac{du_1}{u_1 \ln \frac{1}{u_1}}$$

$$-e^{-1/t} \int_{e^{-e^{1/t}}}^1 K(u_1, f; A_0, A_1) \frac{du_1}{u_1}$$

$$(A_0, A_{-2})_{0,1;K} = \{f : \int_0^1 K(s, f; A_0, A_{-2}) \frac{ds}{s} < \infty\}$$

$$= \{f : \int_0^{e^{-1}} K(s, f; A_0, A_1) d\ln(\ln\frac{1}{s}) < \infty\}$$

More generally,

$$A_{-n} = (A_0, A_{-(n-1)})_{0,1;K}$$

$$= \{f : \int_0^{e_n} K(u, f, A_0, A_1) d \underbrace{\ln(.....(\ln(\frac{1}{u})))}_{n \ times} < \infty\}$$

where $e_n = \exp(-\exp(\exp(...\exp(1))....))$.

Example 72 *In the case of* (L^1, L^∞) *we obtain, by integration by parts:*

$$(L^1, L^\infty)_{(-n)} = \{f : \int_0^{e_n} f^*(s) \underbrace{\ln(.....(\ln(\frac{1}{u})))}_{n \ times} ds < \infty\}$$

that is

$$(L^1, L^\infty)_{(-n)} = L \underbrace{Log(.....LogL)}_{n \ times}$$

The corresponding operators $\Omega_K = \Omega^{(-n)}$ *associated with the pairs* (A_0, A_{-n}) *are given by:*

$$\Omega^{(-n)} f = f(x) \underbrace{\ln(\ln(\ln(...(\ln (r_f(x)))..)}_{n+1 \ times}$$

The commutator theorem in this setting gives: if T is an operator bounded on the pair (A_0, A_{-n}), then, we have, for $n = 0, 1.....$

$$[T, \Omega^{(-n)}] : (A_0, A_{-n})_{\theta,q;K} \rightarrow (A_0, A_{-n})_{\theta,q;K}.$$

For other variants and applications of these theorems we refer to [45].

Let us now explicitly compute the interpolation spaces involved. Using [43], we have,

$$(A_0, A_{-n})_{\theta,q;K} = \{f : \left\{\int_0^1 [(\log \tfrac{1}{t})^{\theta - \frac{1}{q}} K(t, f, A_0, A_{-(n-1)})]^q \tfrac{dt}{t}\right\}^{\frac{1}{q}} < \infty\}$$

Therefore, using (7.27) we are able to compute the corresponding spaces. The easier cases to deal with are when $q = 1$, or $\theta = \frac{1}{q}$.

$$(A_0, A_{-n})_{\theta,1;K} = \{f : \int_0^1 \left(\log \frac{1}{t}\right)^{\theta - 1} K(t, f, A_0, A_{-(n-1)})\frac{dt}{t} < \infty\}$$

$$(A_0, A_{-n})_{\frac{1}{q},q;K} = (A_0, A_{-(n-1)})_{0,q;K}$$

Example 73 *Consider the pair* (L^1, L^∞), *then for* $0 < \theta < 1$,

$$(L^1, L \underbrace{Log(......(LogL))}_{n\ times})_{\theta,1;K} = L \underbrace{Log^\theta(Log(.....(LogL))}_{n\ times}$$

In the same vein we can develop a formulae for the pair $(A_0, \bar{A}_{0,p;K})$.

We consider the higher order operators associated with extrapolation spaces. Again for simplicity assume that $\bar{A}$ is an ordered pair, then the operators Ω_n associated with this pair can be defined by

$$\Omega_n f = \frac{1}{(n-1)!} \int_0^1 D_K(s, A_0, A_1) f (\log s)^{n-1} \frac{ds}{s}$$

The corresponding operators associated with a pair (A_0, A_{-m}), are then given by

$$\Omega_n^{(-m)} f = \frac{1}{(n-1)!} \int_0^1 D_K(e^{-e^{-e^{\cdots^{-e^{\frac{1}{s}}}}}}, A_0, A_1) f (\log s)^{n-1} \frac{ds}{s}$$

Example 74 *For the pair* $(L^1, (L^1, L^\infty)_{-m})$ *we have,*

$$\Omega_n^{(-m)} f = f \underbrace{\log^n(\log(\log ....(\log r_f(x)....)}_{m+1}$$

7.4 Other Operators Ω

In the applications of the theory it is useful to consider other variants of the operators Ω discussed above. For example, in some important examples the "gain" that one obtains by the cancellation effect of $[.,.]$ is so subtle that it manifests itself only at the level of the constants appearing in the corresponding inequalities. This is precisely the case in the following result by Iwaniec-Sbordone (cf. [55])

Theorem 75 *(cf. [55]) Let $1 < r_i < \infty, i = 1, 2$, $r \in [r_1, r_2]$ and suppose that $T : L^r(\Omega, E) \to L^r(\Omega, E)$, where E is a Hilbert space. Then $\forall \varepsilon$ such that $\frac{r}{r_2} - 1 \le \varepsilon \le \frac{r}{r_1} - 1$, we have*

$$\|TS_\varepsilon - S_\varepsilon T\|_{\frac{r}{1+\varepsilon}} \le c_r |\varepsilon| \|f\|_r \tag{7.28}$$

where $S_\varepsilon f = \left(\frac{|f|}{\|f\|_r}\right)^\varepsilon f$, and c_r is independent of f.

Letting $\varepsilon \to 0$ in (7.28) we recover the Rochberg-Weiss commutator theorem

$$\|T\Omega - \Omega T\|_r \le c_r \|f\|_r$$

with $\Omega f = f \log |f|$.

Using our theory we can give a considerable extension of this result. In this section we develop this point in detail. The reader should also consult [55] to compare the methods of proof.

Let $\alpha \in (-1, 1)$, $\alpha \ne 0$, and define

$$\Omega_{E,\alpha} a = \Omega_\alpha a = \alpha \left(\int_1^\infty D_E(t) a \, t^\alpha \frac{dt}{t} - \int_0^1 (I - D_E(t)) a \, t^\alpha \frac{dt}{t} \right)$$

Theorem 76 *(cf. [76]). Let $\bar{A}$ and $\bar{B}$ be a Banach pairs, $T : \bar{A} \to \bar{B}$ be a bounded operator, then, there exists a constant $c > 0$ such that, if $\theta + \alpha > 0$,*

$$\|[\Omega_\alpha, T]f\|_{(B_0, B_1)_{\theta/(\alpha+1), q; E}} \le \frac{c}{\theta} |\alpha| (2c_\alpha)^{\theta/(\alpha+1)} (\alpha+1)^{1/q} \|f\|_{(A_0, A_1)_{\theta+\alpha, q; E}}$$

Proof.   It is easy to see that, according to our definitions, for any Banach pair $\bar{H}$, and for $t > 0$, we have

$$\Omega_{\alpha,\bar{H}}a + a\phi_\alpha(t) =$$

$$= \alpha(\int_t^\infty D_{E,\bar{H}}(s)a\, s^\alpha \frac{ds}{s} - \int_0^t (I - D_{E,\bar{H}}(s))a\, s^\alpha \frac{ds}{s}) \qquad (7.29)$$

where $\phi_\alpha(t) = 1 - t^\alpha$.

Let $\tilde{a}_1(t) = \alpha(\int_0^t (I - D_E(s))a\, s^\alpha \frac{ds}{s})$, then

$$\|\tilde{a}_1(t)\|_{H_1} \leq |\alpha|\, (\int_0^t \|(I - D_E(s))a\|_{H_1}\, s^\alpha \frac{ds}{s})$$

$$\leq \frac{|\alpha|}{(\alpha+1)} t^{\alpha+1} \qquad (7.30)$$

Thus, letting $c_\alpha = [|\alpha|\,(\alpha+1)^{-1}]$ and combining (7.29), (7.30), and (7.7), we get

$$E(c_\alpha t^{\alpha+1}, \Omega_\alpha a + \phi_\alpha(t)a; \bar{H}) \leq |\alpha|\,(\int_t^\infty E(s,a,\bar{H})\, s^\alpha \frac{ds}{s}) \qquad (7.31)$$

Therefore, if $T : \bar{A} \to \bar{B}$, then we can estimate $E(2c_\alpha t^{\alpha+1}, \Omega_{\alpha,\bar{B}}Ta - T\Omega_{\alpha,\bar{A}}a; \bar{B})$ as less or equal than

$$E(c_\alpha t^{\alpha+1}, \Omega_{\alpha,\bar{B}}Ta + \phi_\alpha(t)a; \bar{B}) + E(c_\alpha t^{\alpha+1}, T(\Omega_{\alpha,\bar{A}}a + \phi_\alpha(t)a); \bar{B})$$

Using the fact that T is bounded, and applying (7.31) to each of these terms we get

$$E(2c_\alpha t^{\alpha+1}, \Omega_{\alpha,\bar{B}}Ta - T\Omega_{\alpha,\bar{A}}a; \bar{B}) \leq c|\alpha|\,(\int_t^\infty E(s,a,\bar{A})\, s^\alpha \frac{ds}{s}),$$

where c depends only on the norm of T on the initial pair. An application of Hardy's inequality now yields

$$\{\int_0^\infty [t^\theta E(2c_\alpha t^{\alpha+1}, \Omega_{\alpha,\bar{B}}Ta - T\Omega_{\alpha,\bar{A}}a; \bar{B})]^q \frac{dt}{t}\}^{1/q}$$

$$\leq \frac{c|\alpha|}{\theta}\{\int_0^\infty [E(s,a,\bar{A})s^{\alpha+\theta}]^q \frac{ds}{s}\}^{1/q}$$

and therefore we finally get

$$\|[\Omega_\alpha, T]a\|_{(B_0,B_1)_{\theta/(\alpha+1),q;E}} \leq \frac{c\,|\alpha|}{\theta}(2c_\alpha)^{\theta/(\alpha+1)}(\alpha+1)^{1/q}\|a\|_{(A_0,A_1)_{\theta+\alpha,q;E}}$$

as we wished to show. $\square$

We consider now in detail the special case of L^p spaces. The E functional for the pair (L^1, L^∞) is easy to compute (cf. [56] and Example 61)

$$E(t, f, L^1, L^\infty) = \int_t^\infty \lambda_f(s)ds \qquad (7.32)$$

The interpolation spaces for the pair (L^1, L^∞), by the E method, can be determined using this formula. In fact, if we recall

$$\|f\|_p = \{p \int_0^\infty \lambda_f(s)s^{p-1}ds\}^{1/p}$$

we see, using (7.32), and integration by parts, that

$$\|f\|_{(L^1,L^\infty)_{p-1,1;E}} = [(p-1)p]^{-1}\|f\|_p^p$$

A calculation using Example 61 gives

$$\Omega_\alpha f = f\,|f|^\alpha - f$$

Let us set $S_\alpha f = f\,|f|^\alpha$, then we clearly have $[T, \Omega_\alpha] = [T, S_\alpha]$. Now to apply Theorem 76 we let

$$\frac{\theta}{\alpha+1} = \frac{r}{1+\alpha} - 1, \text{ then } \theta + \alpha = r - 1.$$

Then, the previous discussion gives

$$\|TS_\alpha f - S_\alpha Tf\|_{\frac{r}{1+\alpha}}^{\frac{r}{1+\alpha}} \leq c2^{\frac{r-1-\alpha}{\alpha+1}}\left(\frac{|\alpha|}{(\alpha+1)}\right)^{\frac{r}{\alpha+1}}\frac{1}{(r-1)}\|f\|_r^r$$

Raising both members of the previous inequality to the power $\frac{1+\alpha}{r}$ gives an estimate of Iwaniec- Sbordone type,

$$\left\|T\left(\frac{|f|^\alpha f}{\|f\|_r^\alpha}\right) - \frac{|Tf|^\alpha Tf}{\|f\|_r^\alpha}\right\|_{\frac{r}{1+\alpha}}$$

$$\leq c \left(\frac{|\alpha|}{(\alpha+1)}\right)\left(\frac{1}{(r-1)}\right)^{\frac{\alpha+1}{r}} \|f\|_r \, . \qquad (7.33)$$

In order to obtain a version of Theorem 75 we argue that

$$\left\|T\left(\frac{|f|^\alpha f}{\|f\|_r^\alpha}\right) - \frac{|Tf|^\alpha Tf}{\|Tf\|_r^\alpha}\right\|_{\frac{r}{1+\alpha}} \leq \left\|T\left(\frac{|f|^\alpha f}{\|f\|_r^\alpha}\right) - \frac{|Tf|^\alpha Tf}{\|f\|_r^\alpha}\right\|_{\frac{r}{1+\alpha}} + \\ \left\|\frac{|Tf|^\alpha Tf}{\|f\|_r^\alpha} - \frac{|Tf|^\alpha Tf}{\|Tf\|_r^\alpha}\right\|_{\frac{r}{1+\alpha}}$$

$$= I + II, \; say.$$

Now, I is controlled by (7.33), while II can be readily computed

$$II = \|Tf\|_r \left|\left(\frac{\|Tf\|_r}{\|f\|_r}\right)^\alpha - 1\right|$$

Let $x = \|Tf\|_r$, $y = \|f\|_r$, $u = y/x$, $\phi(u) = u^{\alpha+1} - u$, and assume, as we may, that $\|T\|_{r \to r} \leq 1$, then $u \in [0,1]$, and we have reduced everything to prove that there exists $c > 0$, such that $\forall u \in [0,1]$

$$|\phi(u)| \leq c|\alpha| \qquad (7.34)$$

We study ϕ using calculus, and we see that (7.34) holds with $c = (\frac{1}{1+\alpha})^{\frac{1+\alpha}{\alpha}}$. We conclude the analysis by observing that the factor $1/(1+\alpha)$ is under control by r_2/r. By collecting estimates we see that we have thus obtained an end point version of Theorem 75 by real methods. By reiteration we may obtain the full result.

In a similar manner we may prove that, if

$$\Omega_{\alpha;K}a = \alpha\left(\int_0^1 D_K(s)a\,s^\alpha \frac{ds}{s} - \int_1^\infty (I - D_K(s))a\,s^\alpha \frac{ds}{s}\right)$$

then, we have

Theorem 77 *Let $\bar{A}, \bar{B}$, be Banach pairs, let $T : \bar{A} \to \bar{B}$, be a bounded operator, and let $\alpha \in (-1,1)/\{0\}, 0 < \theta < 1, 0 < q \leq \infty$, and suppose that $0 < \theta - \alpha < 1$. Then, there exists a constant $c = c(\theta, q) > 0$, such that*

$$\|[T, \Omega_\alpha]f\|_{\bar{B}_{\theta,q;K}} \leq c|\alpha| \, \|f\|_{\bar{A}_{\theta-\alpha,q;K}}$$

7.5 Compensated Compactness

In this section we discuss applications of the theory of commutator estimates to compensated compactness. Specifically, we derive estimates for Jacobians of smooth maps, and use commutator estimates to establish perturbed Hodge type decompositions for vector fields.

We review briefly some basic results concerning the Hodge decomposition. We are trying to decompose a vector field F as

$$F = \nabla u + H,$$

where H is a divergence free vector field, that is $div H = 0$.

This is classically done as follows. Suppose that $F \in L^p(R^n, R^n)$, then select u to be such that $\Delta u = div F$, i.e. by letting

$$\nabla u = KF$$

where K is the matrix operator given by

$$K = - \begin{bmatrix} R_1 \otimes R_1 & R_1 \otimes R_2 & ... & R_1 \otimes R_n \\ . & . & ... & . \\ . & . & ... & . \\ R_n \otimes R_1 & ... & ... & R_n \otimes R_n \end{bmatrix}$$

and the $R_j, j = 1, ... n$, are the Riesz transforms. Therefore, the decomposition we seek is

$$F = KF + (I - K)F$$

Moreover, since the Riesz transforms are bounded operators on L^p, $1 < p < \infty$, we have the right control. For vector fields defined on a smooth domain Ω, a similar result holds, and again ∇u is given by a singular integral.

We now consider operators of the form $\Omega_\phi f = f\phi(f)$, where the (generally non-linear) operator ϕ is selected in such a way that for any bounded operator T on L^p with $p \in (1, \infty)$, the commutator $[T, \Omega_\phi]$ is also a bounded operator. More precisely, this means that there exists a function δ, with $\lim_{x \to 0} \delta(x) = 0$, such that,

$$\|[T, \Omega_\phi] f\|_p \leq \|f\|_p \, \delta(\|f\|_p)$$

We have considered a number of operators of this type in the previous sections. For example, in the Rochberg-Weiss theory we can take $\phi(f)(x) = \log|f(x)|$, and then $\delta(x) = cx$. We can also deal with

$$\phi(f) = (\log|f|)^\alpha, 0 \leq \alpha \leq 1$$

or more generally using the work of Kalton (cf. [61])

$$\phi(f) = (\log|f|)^\alpha |\log\, r_f(|f(x)|)|^\beta, 0 \leq \alpha, \beta \leq \alpha + \beta \leq 1$$

where r_f is the rank function defined in Example 62.

Theorem 78 *(cf. [44], [74]) Let $f : R^n \to R^n$, be a mapping of class $C_0^\infty(R^n, R^n)$, and let Ω_ϕ be a commutator on L^n, and let δ be the function associated to Ω_ϕ as above, then*

$$\int_{R^n} J(x, f)\phi(|Df|)dx \leq c\delta(\||Df|\|_{L^n}) \||Df|\|_{L^n}^{n-1}$$

Proof.   We use the method of [44] and the notation of our previous discussion. Using a standard Hodge decomposition write

$$\Omega_\phi(Df)(x) = Dg(x) + H(x) \tag{7.35}$$

with $g \in W^{1,s}(R^n, R^n)$, and $H = (I - K)\Omega_\phi(Df) \in L^s(R^n, GL(n))$ a divergence free matrix-field, $1 < s < \infty$. Note that since this decomposition is unique we have $(I - K)(Df) = 0$. The operator $I - K$ is bounded on L^p for $1 < p < \infty$, and therefore, by the theory of commutators, we get

$$\|(I - K)\Omega_\phi(Df) - \Omega_\phi((I - K)Df)\|_{L^n} \leq c\delta(\||Df|\|_{L^n})$$

thus,

$$\|H\|_n \leq c\delta(\||Df|\|_{L^n}) \tag{7.36}$$

Using the notation of differential forms, $J(x, f)dx = df_1 \wedge df_2 \wedge ... df_n$, and (7.35) takes the form

$$\phi(Df)df_k = dg_k + h_k, k = 1, .., n \tag{7.37}$$

where the h_k are differential forms of degree one whose coefficients coincide with the entries of the k-th column of H. Computing, using (7.37) for $k = 1$, we get

$$\int_{R^n} J(x,f)\phi(Df(x))dx = \int_{R^n} dg_1 \wedge df_2 \wedge .... \wedge df_n + \int_{R^n} h_1 \wedge df_2 \wedge ... \wedge df_n$$

Given the assumption on the vector field f we see that, by Stokes' theorem,

$$\int_{R^n} dg_1 \wedge df_2 \wedge .... \wedge df_n = 0$$

Moreover, the second integral can be estimated, by Hadamard's inequality and (7.36), as follows

$$\int_{R^n} h_1 \wedge df_2 \wedge ... \wedge df_n \leq \int_{R^n} |H(x)| |Df(x)|^{n-1} dx \leq \||H\||_n \||Df\||_n^{n-1}$$

$$\leq c\delta(\||Df\||_n) \||Df\||_n^{n-1}$$

Combining these estimates the desired result follows. $\square$

The following application of Theorem 75 to Hodge decompositions is important in the study of the integrability properties of the Jacobian transformation, as well as in the study of regularity of variational problems (cf. [55]).

Theorem 79 *Let $B = B(a, R)$ be a ball in R^n and let $f \in W_r^1(R^n)$, $r > 1$. Then, for each $\varepsilon \in (1 - r, 1)$ the vector field $|\nabla f|^{-\varepsilon} \nabla f \in L^{\frac{r}{1-\varepsilon}}(R^n)$ can be decomposed as*

$$|\nabla f(x)|^{-\varepsilon} \nabla f(x) = \nabla g(x) + H(x), \quad a.e. \ x \in B$$

where $g \in W_{\frac{r}{1-\varepsilon}}^1(R^n)$ and $H \in L^{\frac{r}{1-\varepsilon}}(R^n, R^n)$ is divergence free and such that

$$\||H\||_{\frac{r}{1-\varepsilon}} \leq c |\varepsilon| \||\nabla f\||_r^{1-\varepsilon}$$

Proof. The idea of the proof has already been shown in the course of the proof of the previous theorem. Thus, let $\Omega_\varepsilon(f) = f |f|^{-\varepsilon}$, and using the operators defined in the introduction, we define now

$$H = (I - K)\Omega_\varepsilon(Df)$$

An application of Theorem 75 or Theorem 76 now gives

$$\||H\||_{\frac{r}{1-\varepsilon}} \leq c |\varepsilon| \||\nabla f\||_r^{1-\varepsilon}$$

as we wished to show. $\square$

7.6   Relationship to Extrapolation

In this section we show some relationships between the theory of commutators and extrapolation. We derive a general functional calculus for interpolation scales which we then relate to the functional calculus of a positive operator on a Banach space.

Since the methods are similar to those of the previous sections, we only indicate the main lines of the arguments here.

In order to motivate our discussion we consider a K/J inequality associated with $\Omega = \Omega_{1,K}$. For simplicity, suppose that $\bar{A}$ is a mutually closed ordered pair. Let $x \in \Delta(\bar{A})$, then by Minkowski's inequality,

$$\|\Omega x\|_{A_0} = K(1, \Omega x; \bar{A}) \leq \int_0^1 \|D_K(r)x\|_{A_0} \frac{dr}{r} + \int_1^\infty \|(I - D_K(r))x\|_{A_1} \frac{dr}{r}$$

Therefore, using (7.24), and the fact that

$$K(r, x; \bar{A}) \leq \min\{\|x\|_{A_0}, r\|x\|_{A_1}\},$$

we obtain

$$\|\Omega x\|_{A_0} \leq \|x\|_{A_0} \int_0^\infty \min\{1, \frac{1}{r}\} \min\{1, \frac{r\|x\|_{A_1}}{\|x\|_{A_0}}\} \frac{dr}{r}$$

which after a computation gives the K/J inequality

$$\|\Omega x\|_{A_0} \leq \|x\|_{A_0} (1 + \log(\frac{\|x\|_{A_1}}{\|x\|_{A_0}})) \tag{7.38}$$

We shall now associate operators to quasi-concave functions as follows. Let φ be a quasi-concave function defined on R_+ such that $\lim_{t \to \infty} \frac{\varphi(t)}{t} = 0$, $\lim_{t \to 0} \varphi(t) = 0$, then, as we have seen before, φ can be represented by

$$\varphi(t) = \int_0^\infty \min\{1, \frac{t}{r}\} d\mu(r)$$

Associated with φ there is an operator Ω_φ defined by

$$\Omega_\varphi(f) = \int_0^1 D_K(r)f \, d\mu(r) - \int_1^\infty (I - D_K(r))f \, d\mu(r) \tag{7.39}$$

$$= \int_0^t D_K(r)f \; d\mu(r) - \int_t^\infty (I - D_K(r))f \; d\mu(r) - f\Phi(t)$$

where $\Phi(t) = \int_1^t d\mu(r)$

We also have the corresponding Calderón operators associated with φ, defined, for measurable functions, by

$$C_\varphi(g)(t) = \int_0^\infty \min\{1, \frac{t}{r}\}g(r)d\mu(r) \qquad (7.40)$$

Observe that (7.39) implies that

$$K(t, \Omega_\varphi f + \Phi(t)f) \leq \int\limits_0^t K(t, D_K(r)f)d\mu(r) + \\ \int\limits_t^\infty K(t, (I - D_K(r))f)d\mu(r)$$

$$K(t, \Omega_\varphi f + \Phi(t)f) \leq C_\varphi(K(.,f))(t). \qquad (7.41)$$

Let ψ be a *function space* on R, a "parameter" for the real method, and let $\bar{A}_{\psi;K}$ denote the interpolation space defined by the condition $K(t, x; \bar{A}) \in \psi$. Let us say that a concave function φ is admissible for $\bar{A}_{\psi;K}$ if the operator C_φ is bounded on ψ. Using the methods of the previous section we get,

Theorem 80 *Suppose that φ is admissible for $\bar{A}_{\psi;K}$, and let T be a bounded operator,$T : \bar{A} \to \bar{A}$, Then the commutator $[T, \Omega_\varphi] = T\Omega_\varphi - \Omega_\varphi T$ is also bounded in $\bar{A}_{\psi;K}$.*

Now we consider the operators Ω_φ in relationship with extrapolation. Let

$$D_\psi(\Omega_\varphi)(\bar{A}) = \{x : x \in \bar{A}_{\psi;K}, \; \Omega_\varphi x \in \bar{A}_{\psi;K}\}$$

with the natural quasi-norm. From the above discussion we see that

$$\left| K(t, \Omega_\varphi x; \bar{A})) - |\Phi(t)| \, K(t, x; \bar{A}) \right| \leq cC_\varphi(K(.,f))(t) \qquad (7.42)$$

Therefore, we readily get

Theorem 81 *Let ψ be admissible for $\bar{A}_{\psi;K}$, then*

$$D_\psi(\Omega_\varphi)(\bar{A}) \equiv \{x : K(t, x; \bar{A}) \, |\Phi(t)| \in \psi\}.$$

Let us now show a natural K/J estimate for Ω_φ

Theorem 82 *Let $\bar{A}$ be a mutually closed, ordered pair, then for $x \in \Delta(\bar{A})$ we have*

$$\|\Omega_\varphi x\|_{A_0} \leq c\|x\|_{A_0}\varphi(\frac{\|x\|_{A_1}}{\|x\|_{A_0}})$$

Proof.   Let $t = 1$ in (7.42), then

$$\|\Omega_\varphi x\|_{A_0} \leq \int_0^\infty \min\{1, \frac{1}{r}\}K(r, x)d\mu(r)$$

$$\leq c\int_0^\infty \min\{1, \frac{1}{r}\}\min\{\|x\|_{A_0}, r\|x\|_{A_1}\}d\mu(r)$$

$$\leq c\|x\|_{A_0}\int_0^\infty \min\{1, \frac{\|x\|_{A_1}}{r\|x\|_{A_0}}\}d\mu(r) = c\|x\|_{A_0}\varphi(\frac{\|x\|_{A_1}}{\|x\|_{A_0}})$$

$\square$

This last estimate persists at the level of commutators

Theorem 83 *Let T be a bounded operator $T : \bar{A} \to \bar{A}$, where $\bar{A}$ is an ordered, mutually closed pair. Then,*

$$\|[T, \Omega_\varphi]x\|_{A_0} \leq c\|x\|_{A_0}\varphi(\frac{\|x\|_{A_1}}{\|x\|_{A_0}})$$

Proof.   We readily get

$$K(t, T\Omega_\varphi x - \Omega_\varphi Tx; \bar{A})) \leq C_\varphi(K(., x))(t)$$

Now, letting $t = 1$, we arrive to

$$\|[T, \Omega_\varphi]x\|_{A_0} \leq cC_\varphi(K(., x; \bar{A})))(1)$$

and the computation on Theorem 82 finishes the proof. $\square$

We should interpret Theorem 83 as an extrapolation theorem for the nonlinear operator $[T, \Omega_\varphi]$ which is in general not bounded on A_0. Because the commutator is not in general a linear operator we cannot extrapolate directly.

We consider in detail the results for Ω. We proceed directly, as in the proof of Theorem 83 and arrive to

$$\|[T,\Omega]f\|_{A_0} \leq c \int_0^\infty \min\{1,\frac{1}{r}\}K(r,f;\bar{A})\frac{dr}{r}$$

$$\leq c[\int_0^1 K(r,f;\bar{A}))\frac{dr}{r} + \int_1^\infty K(r,f)\frac{dr}{r^2}]$$

$$\leq c[\|f\|_{(1)} + \|f\|_{A_0}]$$

Thus, we see that and we have proved the following

$$[T,\Omega] : \bar{A}_{(1)} \to A_0$$

The general case, for a general φ, can be now obtained in a familiar fashion and we shall skip the details.

7.7 A Functional Calculus

One could view the theory of Ω_φ operators as a possible way of constructing a general functional calculus on real interpolation scales. In specific applications we could use this theory to see, for example, which functions *operate* on certain function spaces. We shall presently show that this general functional calculus is in fact closely related to the functional calculus associated with positive operators on a Banach space. The commutator theorems that result are then expressed, and classified, in terms of the growth of the concave functions of the operator. One can expect such results to have applications in the theory of abstract parabolic equations.

In our theory the representation that we use for a quasi-concave function φ plays an important auxiliary role. For other applications, however, we need sometimes to select different representations, which of course should not affect the final estimates. An example of this situation occurs in the functional calculus of positive operator in a Banach space, which we now consider in detail.

Let T be an operator acting on a (complex) Banach space X;. T is positive if $\forall t \geq 0$, the resolvent $R(t) = (T + tI)^{-1}$ exists, and moreover, there exists a constant $c > 0$ such that

$$\|R(t)\| \leq c(1 + t)^{-1}$$

We consider quasi-concave functions represented by:

$$\varphi(t) = \int_0^\infty t(ts+1)^{-1}d\sigma(s) \tag{7.43}$$

where $d\sigma(s)$ is a measure such that $\int_0^\infty (1+s)^{-1}d\sigma(s) < \infty$.
Then, one defines the bounded operator (cf. [87])

$$\varphi(T^{-1}) = \int_0^\infty R(s)d\sigma(s)$$

Associated with a quasi-concave function we have two other quasi-concave functions: $\varphi^*(t) = \frac{t}{\varphi(t)}$, and $\hat\varphi(t) = \frac{1}{\varphi(\frac{1}{t})}$. The identity $\varphi(t)\varphi^*(t) = t$, and the positivity of T, allows us to form the operator $\varphi(T^{-1})^{-1}$, and we are thus lead to define following [87], $\varphi(T) = \hat\varphi(T^{-1})^{-1}$, and finally

$$\varphi(T) = T\hat\varphi^*(T^{-1}) \tag{7.44}$$

The concave function $\hat\varphi^*(t)$ has the representation

$$\hat\varphi^*(t) = \int_0^\infty t(s+t)^{-1}d\sigma(s) \approx \int_0^\infty \min\{1, \frac{t}{s}\}d\sigma(s)$$

and computing $\hat\varphi^*(t^{-1})$ we arrive to the formula

$$\varphi(T)x = \int_0^\infty sT(Ts+I)^{-1}x \, \frac{d\sigma(s)}{s} \, , x \in D(T)$$

The correlation with optimal decompositions is given by the fact that for the ordered pair $(X, D(T))$ the optimal K decompositions are given by

$$D_K(t)x = tT(tT+I)^{-1}x$$

$$(I - D_K(t))x = (tT+I)^{-1}x.$$

In this way, we see that

$$\varphi(T)x = \int_0^\infty D_K(t)x \, \frac{d\sigma(s)}{s}$$

Theorem 84 *Let φ be the quasi-concave function represented by (7.43). Let $d\mu(s) = \frac{d\sigma(s)}{s}$, and define*

$$\Gamma(t) = \int_0^\infty \min\{1, \frac{t}{s}\} d\mu(s)$$

Then,

$$\varphi(T)x - \Omega_\Gamma x = c\,x$$

where Ω_Γ denotes the operator associated with the pair $(X, D(T))$.

Proof. Since $\int_0^\infty \min\{1, \frac{1}{s}\} d\sigma(s) \approx \int_0^\infty (1+s)^{-1} d\sigma(s) < \infty$, we see that $\int_1^\infty \frac{d\sigma(s)}{s} = c < \infty$. Now,

$$\Omega_\Gamma x = \int_0^1 D_K(t)x \frac{d\sigma(t)}{t} - \int_1^\infty (I - D_K(t))x \frac{d\sigma(t)}{t}$$

So that formally, we have,

$$\varphi(T)x - \Omega_\Gamma x = x \int_1^\infty \frac{d\sigma(t)}{t} = cx$$

as we wished to show. □

The point is that, under the notation of the previous theorem, we have, for an operator H acting on the pair $(X, D(T))$,

$$[H, \varphi(T)]x = [H, \varphi(T) - cI]x = [H, \Omega_\Gamma]x$$

Thus, our commutator theorems imply that we have control of the commutators formed between H and concave functions of a positive operator. There is, moreover, an explicit relationship between the growth of the concave function of the operator T, and the type of interpolation scale we use to control the commutators.

The K/J inequality of Theorem 82 in this setting takes the form of a "moment inequality"

$$\|\varphi(T)x\| \le c\|x\|\varphi(\frac{\|Tx\|}{\|x\|})$$

These inequalities are well known in the case $\varphi(s) = s^\theta$, $\theta \in (0, 1)$ (*cf.* Krasnoselski and Sobolevskii [66], for logarithms see Sobolevskii

[97], and for the general case of concave functions we refer to Pustil-nik (cf. [87], corollary 6)). The extrapolation theorem giving the commutation relationship for Ω_φ seems to be new even in this classical setting.

Conversely, once this connection is understood one can see that many aspects of the theory developed in [87] (and the references given there) can be naturally extended to the general setting of the functional calculus associated with optimal decompositions in abstract interpolation theory. This includes the observation that the comparison theory for concave functions acting on a positive operator can be extended to the general setting of optimal decompositions and operators Ω_{φ_1}, Ω_{φ_2}.

In conclusion we should also point out that the relationship between the functional calculus for positive concave functions of a positive operator and the operators Ω had been conjectured earlier in [60]

7.8   A Comment on Calderón Commutators

There are other types of operators associated with optimal decompositions for which we can obtain commutator theorems. The motivation for what follows is the formula for the Calderón commutator (cf. [21])

$$T = c\left\{ \int_0^\infty Q_t M_a P_t \frac{dt}{t} + \int_0^\infty P_t M_a Q_t \frac{dt}{t} \right\}$$

where $a \in L^\infty$, M_a is multiplication by a, and Q_t and P_t are multipliers defined by

$$P_t = I(I + t^2 D^2)^{-1} \text{ and } Q_t = tD(I + t^2 D^2)^{-1}$$

where D is on the Fourier side multiplication by ξ. Observe that if $f \in L^2$, we have

$$f = c \int_0^\infty Q_t^2 f \frac{dt}{t} \tag{7.45}$$

It is not difficult to see that (7.45) is a nearly optimal decomposition of f with respect to the pair of Sobolev spaces (W_2^{-1}, W_2^1); i.e.

for this pair we can choose $D_J(t)f = Q_t^2$. One also notices that P_t is related to Q_t^2 through the usual relation between the $D_K(t)$ and $D_J(t)$ nearly optimal decompositions:

$$t\frac{d}{dt}P_t = cQ_t^2$$

(with $c = -2$, in this case).

This leads to the question of how much of the analysis of [21] can be pushed to this abstract setting (and conversely!). For example, in the former direction, R. Rochberg and the author have shown, and this is not hard, that if $\bar{A}$ is a Banach pair, and $T : \bar{A} \to \bar{A}$, is a bounded operator then the operator

$$\Gamma f = \int_0^\infty D_J(t) T D_J(t) f \frac{dt}{t}$$

is bounded on the real interpolation spaces $\bar{A}_{\theta,q}$.

7.9   Notes and Further Results

In [69], Li, McIntosh and Zhang provide another approach to estimates for Jacobians through the use of weighted norm inequalities. A general overview of recent results in nonlinear partial differential equations, using the methods of real harmonic analysis, can be found in Müller's survey [83].

Although commutators of a bounded operator on an interpolation scale with the Ω_n operators are not necessarily bounded, there is some cancellation which allows for the following type of results considered in [77]. Let $0 < \theta < 1, 1 \leq q \leq \infty, \alpha \in R$, and define

$$\phi_{\theta,\log^\alpha,q}(f) = \left\{ \int_0^\infty \left| t^{-\theta} f(t)(1 + |\log t|)^\alpha \right|^q \frac{dt}{t} \right\}^{\frac{1}{q}}$$

Define interpolation spaces $\bar{A}_{\theta,\log^\alpha,q;K}, 0 < \theta < 1, 0 < q \leq \infty$, by

$$\bar{A}_{\theta,\log^\alpha,q;K} = \{ a : \|a\|_{\bar{A}_{\theta,\log^\alpha,q;K}} = \{a/\phi_{\theta,\log^\alpha,q} \left(K(., a, \bar{A}) \right) < \infty \}$$

It is not hard to see that, if $T : \bar{A} \to \bar{B}$ is a bounded operator, then

$$K(t, [T, \Omega_n]f; \bar{B}) \le c \int_0^\infty |\log s|^{n-1} \min\{1, \frac{t}{s}\} K(s, f; \bar{A}) \frac{ds}{s}$$

Therefore, we obtain

$$\|[T, \Omega_n]f\|_{\bar{B}_{\theta,q;K}} \le c \|f\|_{\bar{A}_{\theta, \log^{n-1}, q; K}}$$

Observe that the case $n = 1$ corresponds to Theorem 67. See [77] for the details.

The results of Section **7.5**, were motivated, in part, by a joint project with Professor L. Caffarelli on the real Monge Ampere equation and Brennier rearrangements. It was Professor Caffarelli who suggested to me the study of the theory of compensated compactness, and Hodge decompositions, with the hope that we could adapt parts of it to our problem. The later part of that project I still have to accomplish.

The relationship between optimal decompositions and a functional calculus for a positive operator can be extended in several directions. For example it would be of interest to relate our work here with the recent results of Dore [39], and Cowling, Doust, McIntosh, and Yagi [23].

Chapter 8

Sobolev Imbedding Theorems and Extrapolation of Infinitely Many Operators

The Sobolev embedding theorems play an important role in these notes as a source of examples and applications. In this brief chapter we take a different tack we model from a form of the Sobolev embedding theorem an abstract interpolation theorem that suggests the possibility of a more general theory of interpolation. The relationship between these results and the theory associated with the operators Ω_φ of Chapter 7 ought to be investigated. In forthcoming work [78] we show that this set up can be used to study logarithmic Sobolev inequalities.

8.1 Averages of Operators

From the abstract point of view the general question is as follows. Suppose that we have a family of operators $T = \{T(t)\}_{t \in (0,\infty)}$, each acting on a Banach pair $\bar{A}$; then under what conditions can we assert that an appropriate average of this family is bounded on intermediate spaces? There are many possible variations on this general question.

Let us start by saying what types of "averages" we have in mind:

$$T = average(T(t)) = \int_0^\infty T(t)d\mu(t)$$

in a suitable sense. For example if $d\mu(t)$ is a probability measure, and $T(t) = T$, $\forall T$, then the answer is trivial: T is bounded on the interpolation spaces for the pair $\bar{A}$. Another trivial result is: if $\int_0^\infty m_{\theta,q}(t)d\mu(t) < \infty$, where $m_{\theta,q}(t) = \|T(t)\|_{\bar{A}_{\theta,q}\to\bar{A}_{\theta,q}}$ then average(T) is bounded on $\bar{A}_{\theta,q}$.

The next result is stated in such a way as to make it easy to compare directly with an abstract version of the Sobolev embedding theorem. This remark should explain our somewhat awkward choice of parameters.

Theorem 85 *Let $\bar{A}$, $\bar{B}$, be Banach pairs, and let $\{T(t)\}_{t\in(0,\infty)}$ be a continuous family of contractions, $T(t) : \bar{A} \to \bar{B}$, such that for some $\mu > 1$, we have*

$$\|T(t)f\|_{B_1} \le ct^{-\mu/2}\|f\|_{A_0}$$

Then, if we let formally,

$$T^{-1/2} = \int_0^\infty T(t)t^{-1/2}dt$$

we have, $\forall\theta$ such that $(1 - \theta)\mu > 1$,

$$T^{-1/2} : \bar{A}_{\theta,q} \to \bar{B}_{\gamma,\infty}$$

$\forall\theta$ such that $(1-\theta)\mu > 1$, $\gamma = [1-(1-\theta)^{-1}\mu^{-1}]\theta+(1-\theta)^{-1}\mu^{-1}$.

Proof. Let $f \in \bar{A}_{\theta,q}$, we estimate the K functional, $K(t,T^{-1/2}f,\bar{B}_{\theta,q}, B_1)$. For $M > 0$, to be specified later we split

$$T^{-1/2}f = \int_0^M t^{-1/2}T(t)f \; dt + \int_M^\infty t^{-1/2}T(t)f \; dt = g_0 + g_1$$

and estimate each of these terms. Let us observe first that from $T(t) : A_1 \to B_1$, $T(t) : \bar{A}_0 \to B_1$, the estimates we have available and interpolation we have

$$\|T(t)\|_{\bar{A}_{\theta,q}\to B_1} \le c_{\theta,q}t^{-\mu(1-\theta)/2}$$

Consequently,

$$\|g_1\|_{B_1} \leq c\|f\|_{\bar{A}_{\theta,q}} \int_M^\infty t^{-1/2} t^{-\mu(1-\theta)/2} dt$$

$$\leq c\|f\|_{\bar{A}_{\theta,q}} M^{\frac{1}{2} - \frac{\mu(1-\theta)}{2}}$$

To estimate $\|g_0\|_{\bar{A}_{\theta,q}}$ we use the fact that $\|T(t)\|_{\bar{A}_{\theta,q} \to \bar{B}_{\theta,q}} \leq c$, and obtain

$$\|g_0\|_{B_{\theta,q}} \leq c M^{\frac{1}{2}} \|f\|_{\bar{A}_{\theta,q}}$$

Therefore, if we choose $M = t^{\frac{1}{\mu(1-\theta)}}$, we obtain

$$K(t, T^{-1/2}f, \bar{B}_{\theta,q}, B_1) \leq \|g_0\|_{\bar{A}_{\theta,q}} + t\|g_1\|_{B_1}$$

$$\leq c\, t^{\frac{1}{\mu(1-\theta)}} \|f\|_{\bar{A}_{\theta,q}}$$

In other words, we have shown that

$$T^{-1/2} : \bar{A}_{\theta,q} \to (\bar{B}_{\theta,q}, B_1)_{\frac{1}{\mu(1-\theta)}, \infty} = \bar{B}_{\gamma,\infty}$$

with γ defined as in the statement of the theorem. $\square$

Corollary 86 *Suppose that $\mu > 2$, then under the assumptions of Theorem 85 we have*

$$T^{-1/2} : \bar{A}_{1/2,2} \to \bar{B}_{1/\mu+1/2,2}$$

Proof.   We use the reiteration theorem. Let θ_0, θ_1, be such that $\theta_0 > 1/2$, $\theta_1 < 1/2$, and $1/(1 - \theta_i) < \mu$, $\beta \in (0,1)$ such that $(1-\beta)\theta_0 + \beta\theta_1 = 1/2$, then the γ_0, γ_1 corresponding to the θ_i satisfy $(1-\beta)\gamma_0 + \beta\gamma_1 = 1/\mu + 1/2$. $\square$

It is not difficult to write down a rather more general version of this result for $T^{-\alpha}$ but we shall not pursue this here. We do turn however to the Sobolev embedding theorem of Varopoulos [102] that was our original motivation.

Example 87 *(cf [102]) Let e^{-Ht} be a symmetric Markov semigroup on $L^2(\Omega)$, and suppose that for $\mu > 2$ we have*

$$\|e^{-Ht}f\|_{L^\infty} \leq c\, t^{-\frac{\mu}{4}} \|f\|_{L^2}$$

Then, the fractional power, $H^{-1/2} = \int_0^\infty t^{-1/2} e^{-Ht} dt$, is a bounded operator

$$H^{-1/2} : L^2 \to L^{\frac{2\mu}{\mu-2}}$$

Proof.   The assumption on $\{e^{-Ht}\}$ implies that

$$\|e^{-Ht}f\|_{L^\infty} \le c\, t^{-\frac{\mu}{2}}\|f\|_{L^1}$$

In fact, by duality $e^{-Ht} : L^1 \to L^2$, with norm smaller than or equal than $ct^{-\frac{\mu}{4}}$, then, since $e^{-Ht} = e^{-Ht/2}e^{-Ht/2}$, we get

$$\|e^{-Ht}f\|_{L^\infty} \le \|e^{\frac{-Ht}{2}}\|_{L^1\to L^2}\|e^{\frac{-Ht}{2}}\|_{L^2\to L^\infty}\|f\|_{L^1} \le ct^{-\frac{\mu}{2}}\|f\|_{L^1}$$

Therefore, if we apply Theorem 86 we obtain:

$$H^{-1/2} : L^2 = (L^1, L^\infty)_{\frac{1}{2},2} \to (L^1, L^\infty)_{\frac{1}{\mu}+\frac{1}{2},2}$$

Now,

$$(L^1, L^\infty)_{\frac{1}{\mu}+\frac{1}{2},2} \subset (L^1, L^\infty)_{\frac{1}{\mu}+\frac{1}{2},r} = L^r, \frac{1}{r} = \frac{1}{2} - \frac{1}{\mu},$$

and the desired result follows. $\Box$

Chapter 9

Some Remarks on Extrapolation Spaces and Abstract Parabolic Equations

Interpolation spaces play an important role in the theory of abstract differential equations in Banach spaces, and there is a substantial literature devoted to this field (cf. [31], [2], and the references therein). Here we shall only briefly begin to explore the role of extrapolation spaces in the theory of parabolic equations in Banach spaces.

We shall consider the limiting cases of the theory of maximal regularity for these equations on L^p spaces (cf. [32], [36]). In order to apply our theory and be able to extrapolate from known estimates in the literature we need to be very careful about the norms that we consider. One could also obtain weaker results through the use of reiteration theorems, as in Chapter 6.

Remark. In a number of papers Amann, and DaPrato and Grisvard (cf. [2], and [33]) have studied a notion of "extrapolation spaces" of an entirely different nature than those considered here.

9.1   Maximal Regularity

Consider the formula for the mild solution of the parabolic initial value problem

$$\begin{cases} u' = A(t)u + f(t) \\ \quad u(0) = u_0 \end{cases} \tag{9.1}$$

where A is the generator of an analytic semigroup $S(t)$, with dense domain D_A, on a Banach space E, and u_0 belongs to an appropriate subspace of E. The solution is formally given by,

$$u(t) = S(t)u_0 + \int_0^t S(t-s)f(s)ds \tag{9.2}$$

Thus, a central problem in the theory is to give a meaning to each of the terms in (9.2). This problem can be often formulated in term of the continuity of the underlying operators

$$Su_0(t) = S(t)u_0, \ \Gamma f(t) = \int_0^t S(t-s)f(s)ds$$

in suitable function spaces. The delicate part of the problem is to deal with the operator Γ, which should be considered as an abstract singular integral operator.

Interpolation spaces play a key role here since they are the natural ambient for the study of the operators at hand. We need to explain carefully how these spaces are constructed in our setting in order to be able to use estimates available in the literature.

Recall that if A is the generator of an analytic semigroup then, there exist $\theta \in (\frac{\pi}{2}, \pi)$, $c > 0$, such that, if $\arg(z) < \theta$, then $\forall x \in E$,

$$\left\| z(zI - A)^{-1}x \right\|_E \le c \|x\|_E \tag{9.3}$$

It is also well known that there exists a constant $c > 0$, such that the following estimates hold $\forall t > 0, \forall x \in E$,

$$\|S(t)x\|_E \le c\|x\|_E \tag{9.4}$$

$$t\|AS(t)x\|_E \le c\|x\|_E \tag{9.5}$$

Consider D_A with the norm given by

$$\|x\|_{D_A} = \|Ax\|_E$$

Then, it is well known and readily seen (cf. [8], [87]) that

Lemma 88 *Let $R(t) = (A + tI)^{-1}$, then*

$$K(t, x; E, D_A) \approx \|AR(\frac{1}{t})x\|_E \qquad (9.6)$$

Proof. If $x = f_0 + f_1$ is a decomposition with $f_0 \in E$, $f_1 \in D_A$, then

$$\|AR(\frac{1}{t})x\|_E \leq \|AR(\frac{1}{t})f_0\|_E + \|AR(\frac{1}{t})f_1\|_E$$

Now, write $AR(\frac{1}{t})f_0 = f_0 - \frac{1}{t}R(\frac{1}{t})f_0$, then by the triangle inequality, and (9.3), we get

$$\|AR(\frac{1}{t})f_0\|_E \leq c\|f_0\|_E$$

and

$$\|AR(\frac{1}{t})f_1\|_E \leq ct\|Af_1\|_E \leq ct\,\|f\|_{D_A}$$

Therefore,

$$\|AR(\frac{1}{t})x\|_E \leq c\left(\|f_0\|_E + t\,\|f_1\|_{D_A}\right)$$

and taking infimum over all decompositions, we get

$$\|AR(\frac{1}{t})x\|_E \leq cK(t, x; E, D_A)$$

In the opposite direction, write

$$x = AR(\frac{1}{t})x + \frac{1}{t}R(\frac{1}{t})x$$

Then,

$$K(t, x; E, D_A) \leq \left\|AR(\frac{1}{t})x\right\|_E + t\left\|R(\frac{1}{t})x\right\|_{D_A}$$

$$\leq 2\left\|AR(\frac{1}{t})x\right\|_E$$

$\square$

The spaces $(E, D_A)_{\theta, q; K}$, $0 < \theta < 1, 1 \leq q \leq \infty$, can be thus characterized by

$$(E, D_A)_{\theta, q; K} = \{x : \left\{\int_0^\infty \left(\left\|AR(\frac{1}{t})x\right\|_E t^{-\theta}\right)^q \frac{dt}{t}\right\}^{1/q} < \infty\}$$

The corresponding norms are given by

$$\|x\|_{(E,D_A)_{\theta,q,K}} = c_{\theta,q}\left\{\int_0^\infty \left(\left\|AR(\frac{1}{t})x\right\|_E t^{-\theta}\right)^q \frac{dt}{t}\right\}^{1/q}$$

where $c_{\theta,q} = ((1-\theta)\theta q)^{1/q}$.

For $\theta = 0$, we have

$$(E,D_A)_{\theta,q;K} = \left\{x : \left\{\int_1^\infty (\|AR(t)x\|_E)^q \frac{dt}{t}\right\}^{1/q} < \infty\right\}$$

Observe that compared with the spaces $D_A(\theta, q)$ as defined in [36], we have

$$(E,D_A)_{\theta,q;K} = D_A(1-\theta, q)$$

and

$$\|x\|_{(E,D_A)_{\theta,q;K}} = c_{\theta,q}\|x\|_{D_A(1-\theta,q)}$$

Let us also remark that the pair (E, D_A) is mutually closed. For example, (cf. [13]) we have

$$\lim_{t\to 0}\frac{K(t,x;E,D_A)}{t} \approx \lim_{t\to\infty} t\,\|AR(t)x\|_E = \|x\|_{D_A}$$

Let us now return to the study of the operators S, and Γ. As we indicated above the operator S is not difficult to control, for example from (9.4) and (9.5) we see that

$$t\,\|AS(t)x\|_E \leq cK(t,x;E,D_A) \tag{9.7}$$

Therefore, integrating (9.7) with respect to t, we see that

$$\|Sx\|_{L^1(0,T,D_A)} \leq c\|x\|_{(E,D_A)_{(1)}} \tag{9.8}$$

Let us now illustrate the role of extrapolation spaces showing an end point result for Γ

Theorem 89 *Let $T > 0$, then if $f \in L^1(0,T;(E,D_A)_{(1)})$, we have that $A\Gamma f \in L^1(0,T;E)$*

Proof.   Let us define the operator

$$\Gamma_A f(t) = A(\Gamma f(t))$$

It is shown in [32] and [36] that

$$\Gamma_A : L^1(0, T; (E, D_A)_{\theta,1;K}) \to L^1(0, T; (E, D_A)_{\theta,1;K})$$

with

$$\|\Gamma_A\|_{L^1(0,T;(E,D_A)_{\theta,1;K}) \to L^1(0,T;(E,D_A)_{\theta,1;K})} \le c\theta^{-1}, \text{ as } \theta \to 0$$

(cf. [36], page 72).
    By extrapolation,

$$\Gamma_A : \sum_\theta \left(\frac{1}{\theta} L^1(0, T; (E, D_A)_{\theta,1;K}) \right) \to L^1(0, T; E)$$

It remains to identify the corresponding extrapolation space. Now, since for any pair $\bar{H}$ we have (cf. [8])

$$K(t, f; L^1(0, T; H_0), L^1(0, T; H_1)) \approx \int_0^T K(t, f(x); \bar{H}) dx \quad (9.9)$$

We see that

$$L^1(0, T; (E, D_A)_{\theta,1;K}) = (L^1(0, T; E), L^1(0, T; D_A))_{\theta,1;K}$$

Therefore,

$$\sum_\theta (\frac{1}{\theta} L^1(0, T; (E, D_A)_{\theta,1;K}))$$

$$= \sum_\theta (\frac{1}{\theta} (L^1(0, T; E), L^1(0, T; D_A))_{\theta,1;K})$$

$$= (L^1(0, T; E), L^1(0, T; D_A))_{0,1;K} \text{ (by (2.22))}$$

$$= (L^1(0, T; (E, D_A)_{0,1;K}) \text{ (by (9.9))}$$

and the result follows. $\square$
    As a consequence we have the following regularity result (cf. [36], Theorem 21)

Corollary 90 *Let $f \in (L^1(0,T;(E,D_A)_{0,1;K})$ then $\frac{d}{dt}\Gamma f(t) \in L^1(D_A)$, and moreover*

$$\frac{d}{dt}\Gamma f(t) = A\Gamma f(t) + f$$

One could prove a whole family of end point results for Γ in a similar fashion. For a more complete exploration of applications we refer to work currently in progress (cf. [2])

We conclude this chapter with an example illustrating how extrapolation spaces in this setting can be computed in a concrete situation.

Let L be a second order elliptic operator on R^n, given by

$$Lu = \sum_{1 \le i,j \le n} D_j(a_{ij}(x)D_i u) + \sum_{i=1}^{n} b_i(x)D_i u(x) + c(x)u$$

where $D_i = \frac{\partial}{\partial x_i}$, and the coefficients a_{ij}, b_i and c are uniformly continuous and bounded. Let

$$A : D_A \subset L^1(R^n) \to L^1(R^n)$$

be the operator defined by $Au = Lu$, and where D_A is defined by

$$D_A = \{u \in C_U^2(R^n) \cap L^1(R^n) : Lu \in L^1(R^n)\}$$

where U stands for uniformly continuous and bounded. Moreover, let A_1 be the closure of B in L^1. Then Cannarsa and Vespri [16] have shown that A_1 generates an analytic semigroup. In [37] it is shown that

$$(L^1(R^n), D_{A_1})_{\theta,1;K} = (L^1(R^n), W_1^2(R^n))_{\theta,1;K} \ , \ 0 < \theta < 1, 0 < q \le \infty.$$
$$\tag{9.10}$$

From this result it is easy to derive a characterization of the corresponding extrapolation spaces.

Theorem 91

$$(L^1, D_{A_1})_{0,1;K} = (L^1, W_1^2)_{0,1;K}$$

$$= \{f : \|f\|_{L^1} + \int_0^1 \omega_1(t,f)\frac{dt}{t} < \infty\}$$

where $\omega_1(t,f)$ denotes the L^1 modulus of continuity of f.

Proof. Let $\theta \in (0, \frac{1}{2})$, by (9.10), and the reiteration property (6.1), we have

$$(L^1, D_{A_1})_{0,1;K} = (L^1, (L^1, D_{A_1})_{\theta,1;K})_{0,1;K}$$

$$= (L^1, (L^1, W_1^2)_{\theta,1;K})_{0,1;K}$$

and by reiteration (cf. [101])

$$= (L^1, (L^1, W_1^1)_{\theta',1;K})_{0,1;K}$$

$$= (L^1, W_1^1)_{0,1;K}$$

and the result follows. □

In conclusion we remark that another potentially interesting aspect of the use of extrapolation methods in this area are the constancy properties described in Chapter **6**. The standard assumptions in the theory require either that the spaces $D_{A(t)}$ are constant in which case automatically all the interpolation spaces for the pair $(E, D_{A(t)})$ are constant in t, or directly assuming that suitable interpolation spaces for this pair are constant. As we pointed out in Chapter **6**, for extrapolation methods we need less stringent conditions to guarantee the constancy of the spaces.

Chapter 10

Optimal Decompositions, Scales, and Nash-Moser Iteration

In this chapter we consider the relationship between interpolation and the implicit function theorems usually associated with the names of Kolmogorov-Nash-Arnold-Moser. These results play an important role in many problems of non-linear partial differential equations. We show a connection between optimal decompositions and iterative methods to solve non-linear equations in Banach spaces.

Implicit function theorems have been studied in the very general framework of scales of spaces (cf. [47] and the references quoted therein). There have also been a few contributions using interpolation theory (cf. [68]), but as far as we know there have been no attempts to relate the methods of the theory of scales with real interpolation theory through the use of optimal decompositions.

Our program in this chapter is as follows. First we review Moser's [81] original approach to solve non-linear equations and point out the close relationship between Moser's methods and the real method of interpolation. In the following section we relate the concept of a scale with "smoothing" with optimal K and J decompositions. It turns out that smoothing allows one to produce, by rescaling, optimal decompositions for all interpolation spaces between any two spaces in the scale. We use this idea to identify the ad-hoc spaces that crucially intervene in the Nash-Moser theorem as formulated by

Hormander [49] as those obtained by the $(.,.)_{\theta,\infty}$ method. Finally in the last section we review very briefly Hormander's approach to the Nash-Moser theorem using paracommutators. These operators are directly expressed in terms of optimal decompositions.

10.1   Moser's Approach to Solving Non-Linear Equations

Although the main import of the iterative methods is to solve non-linear equations we start by recasting the simplest possible iteration method of Moser for solving linear equations. A perusal of [81] shows that although this paper was developed independently from the theory of interpolation spaces, which was in a developing stage at the time, there is a close contact between these theories. Indeed, since then other authors have developed this remark in different directions (cf. [68]).

We shall now rephrase the setting of Moser's theory [81] using interpolation theory. Let $\{X_\theta\}_{\theta \in [0,1]}$ be an ordered, decreasing, scale of spaces. We assume, moreover, that the scale behaves almost as an interpolation scale in the sense that the spaces in it satisfy

$$(X_0, X_1)_{\theta,1;J} \subset X_\theta \subset (X_0, X_1)_{\theta,\infty;K} \qquad (10.1)$$

Note that (10.1) will be satisfied if the spaces X_θ, $\theta \in (0,1)$ are constructed by letting

$$X_\theta = \mathcal{F}_\theta(X_0, X_1)$$

with $\{\mathcal{F}_\theta\}_{\theta \in (0,1)}$, a family of exact interpolation functors with each $\mathcal{F}_\theta$ exact of order θ.

Let $L : X_1 \to X_0$ be an operator, and define the K_L functional by:

$$K_L(t, g, X_0, X_1) = \inf_{w \in X_1} \{\|Lw - g\|_{X_0} + t\|w\|_{X_1}\}$$

The spaces $\bar{X}_{\theta,q;K}^{(L)}$ are defined by imitating the classical definition, thus, for example, $g \in (X_0, X_1)_{\theta,\infty;K}^{(L)}$ means that $\exists C > 0$ such that $\forall t \in (0,1)$,

$$K_L(t, g, X_0, X_1) \leq Ct^\theta$$

This can be reformulated as saying that $g \in (X_0, X_r)_{\theta,\infty,K}^{(L)} \iff \forall t \in (0, 1), \exists D_{K_L}(t)g \in X_1$ such that

$$K_L(t, g, X_0, X_r) \approx \|LD_{K_L}(t)g - g\|_{X_0} + t\|D_{K_L}(t)g\|_{X_1} \leq c \, t^\theta \|g\|_{\theta,q;K}^{(L)}$$

This forces, in particular, the convergence of $LD_K(t)g$ to g, as $t \to 0$, at the rate t^θ, while at the same time we have control on the growth of $\|D_{K_L}(t)g\|_{X_1}$. Thus, $\{D_{K_L}(t)g\}$ forms a family of "approximate solutions" to the equation

$$Lf = g$$

The issue then becomes the convergence of $D_{K_L}(t)g$, this necessitates a rescaling of the parameters, *i.e.* the use of other spaces in the scale and what, after discretization, essentially amounts to the J method: that is the study of the convergence of the sequence through the study of the speed of decay of successive differences. Let us plot through the details in the classical case, i.e. when $L = Identity$, to see that in this case the study of the convergence of the family of approximate solutions really amounts to (a weak version of) the fundamental lemma of interpolation theory. Indeed, the argument that Moser gives actually proves that $(X_0, X_1)_{\theta,\infty;K} \subset (X_0, X_1)_{s,1;J} \subset X_s$, if $s < \theta$. In fact this is best understood through discretization. So let $g \in (X_0, X_1)_{\theta,\infty;K}$ and let $t \approx 2^{-n}$, $g_n = D_K(2^{-n})g$, and study the convergence of $g = \sum_{n=1}^\infty g_{n+1} - g_n$. By the triangle inequality

$$\|g_{n+1} - g_n\|_{X_0} \leq c2^{-n\theta}$$

while,

$$\|g_{n+1} - g_n\|_{X_1} \leq c2^{-n(\theta-1)}$$

which implies

$$\|g_{n+1} - g_n\|_{X_s} \leq c(\|g_{n+1} - g_n\|_{X_0})^{1-s}(\|g_{n+1} - g_n\|_{X_1})^{s}$$

$$\|g_{n+1} - g_n\|_{X_s} \leq c2^{-n\theta(1-s)-n(\theta-1)s} = c2^{n(s-\theta)}$$

This geometric decay shows that $\sum_{n=1}^\infty g_{n+1} - g_n$ converges in X_s. In fact the estimates above show that

$$J(2^{-n}, g_{n+1} - g_n; X_0, X_1) \leq 2^{-n\theta}$$

and therefore $g \in (X_0, X_1)_{s,1;J}$.

With suitable assumptions the analysis also works for operators L acting on scales. In particular, let us consider the embedding

$$(X_0, X_1)^{(L)}_{\theta, \infty; K} \subset L((X_0, X_1)_{s,1;J}), \, s < \theta$$

That is, we try to prove that if g is in the space $(X_0, X_1)^{(L)}_{\theta, \infty; K}$ then the equation $Lf = g$, has a solution $f \in (X_0, X_1)_{s,1;J}$. The only change needed is an appropriate condition in order to be able to estimate $\|g_{n+1} - g_n\|_{X_0}$ by $\|Lg_{n+1} - Lg_n\|_{X_0}$, (which of course was not needed in the case $L = identity$). Thus if we assume, for example, that L is an operator such that $L : X_1 \to X_0$, and such that $\exists C > 0$, such that

$$\|x\|_{X_0} \leq C\|Lx\|_{X_0}, \forall x \in X_1$$

the argument above works.

Thus, we should see these results as extensions of the fundamental lemma. This description suggests a number of projects (eg. extend the notion of approximate solution, extend the reiteration argument, extrapolate the end point results, compute the K functionals for specific operators, etc)...

10.2   Scales with Smoothing and Interpolation

We study scales of spaces with an additional structure provided by "smoothing" and we relate them to optimal decompositions for interpolation spaces.

Let us recall the concept of quasilinearizable Banach pair (cf. [86], [8])

Definition 92 *Let $\bar{A}$ be a Banach pair. We say that $\bar{A}$ is quasilinearizable if there exist families of linear operators $V_i(t) : \sum(\bar{A}) \to \Delta(\bar{A})$, $i = 0, 1, t \in R_+$, and moreover $\forall t > 0$, we have,*

$$V_0(t)f + V_1(t)f = f$$

$$\|V_0(t)f\|_{A_0} \leq c\|f\|_{A_0} \, , \, \|V_0(t)f\|_{A_0} \leq ct\|f\|_{A_1}$$

$$t\|V_1(t)f\|_{A_1} \le c\|f\|_{A_0}, t\|V_1(t)f\|_{A_1} \le ct\|f\|_{A_1}$$

where c is an absolute constant

The motivation for this definition is given by the readily verified fact that if $\bar{A}$ is quasilinearizable then

$$K(t, f; \bar{A}) \approx \|V_0(t)f\|_{A_0} + t\|V_1(t)f\|_{A_1}$$

In the case of an ordered pair $\bar{A}$ we only need families of operators $\{V_i(t)\}_{t\in I}$, $i = 0, 1$, satisfying all the conditions above, with $I = (0, 1]$.

We now turn to the setting of the analysis in [50]. Let $\{E^a\}_{a\ge 0}$ be a decreasing family of Banach spaces with injections $E^b \subset E^a$ of norm one, when $b \ge a$. Assume that a family of continuous linear operators $\{S(\theta)\}_{\theta\ge 1}$, is given such that

$$S(\theta) : E^0 \to E^\infty = \bigcap_a E^a, \theta \ge 1$$

$$\|S(\theta)f\|_b \le c\|f\|_a, b \le a \tag{10.2}$$

$$\|S'(\theta)f\|_b \le c\, \theta^{b-a-1}\|f\|_a, \tag{10.3}$$

$$u \in E^0, and\ S(\theta)u \to v\ in\ E^0\ as\ \theta \to \infty \implies v = u \tag{10.4}$$

As a consequence of (10.2), (10.3) and (10.4), we get (cf. [49])

$$\|S(\theta)f\|_b \le c\, \theta^{b-a}\|f\|_a(b - a)^{-1}, a < b \tag{10.5}$$

$$\|(I - S(\theta))f\|_b \le c\, \theta^{b-a}\|f\|_a(a - b)^{-1}, a > b \tag{10.6}$$

A consequence of (10.5) and (10.6) is the logarithmic convexity of the norms of the spaces in the scale: (cf. [49])

$$\|f\|_c \le c_\lambda \|f\|_a^\lambda \|f\|_b^{1-\lambda} \tag{10.7}$$

where $c = \lambda a + (1 - \lambda)b$, $c_\lambda \le [(b - a)\lambda(1 - \lambda)]^{-1}$.

A scale $\{E^a\}_{a\ge 0}$ associated with a family of operators $\{S(\theta)\}_{\theta\ge 1}$ with properties (10.2), (10.3), (10.4), (10.5), (10.6), (10.7), is usually called a *Banach scale with smoothing*.

Associated with $\{E^a\}_{a\ge 0}$ Hormander defines the spaces $\{E_*^a\}_{a\ge 0}$ as follows.

Definition 93 *For a > 0, let $E_*^a = \{u \in E^0 : \exists M > 0$ such that $\|u\|_0 \leq M, \|S'(\theta)u\|_0 \leq M\, \theta^{-a-1}, \|S'(\theta)u\|_{a+1} \leq M, \theta \geq 1\}$ and $\|u\|_a^*$ is the infimum of all the admissible M's.*

Our first observation is that smoothing provides optimal decompositions for all interpolation scales between two fixed spaces in the scale.

Lemma 94 *Let $\{E^a\}_{a\geq 0}$ be a Banach scale with smoothing. Then,*
   (i) For $0 \leq \alpha < \beta$ the pair (E^α, E^β), is quasilinearizable with $V_0(t) = I - S(t^{-\gamma})$, $V_1(t) = S(t^{-\gamma})$, $\gamma = 1/(\beta - \alpha)$, $t \in (0,1]$.
   (ii) For $0 \leq \alpha < \beta$, $0 < \theta < 1$, $0 < q \leq \infty$, $f \in (E^\alpha, E^\beta)_{\theta,\infty;J}$ can be represented by

$$f = S(1)f + (-\gamma) \int_0^1 S'(t^{-\gamma})ft^{-\gamma}\frac{dt}{t}$$

and letting $D_J(t)f = (-\gamma)(t^{-\gamma}S'(t^{-\gamma})f)$

$$\|f\|_{\theta,q;J} \approx \{\int_0^1 [t^{-\theta}J(t, D_J(t)f; E^\alpha, E^\beta)]^q\frac{dt}{t}\}^{1/q}$$

Proof.   *(i) We verify all the conditions of Definition 92:*

$$\|(I - S(t^{-\gamma}))f\|_\alpha \leq 2\|f\|_\alpha \text{ (by the triangle inequality and (10.2))}$$

$$\|(I - S(t^{-\gamma}))f\|_\alpha \leq (\beta - \alpha)^{-1}t^{-1}\|f\|_\beta \text{ (by (10.6))),}$$

$$t\|S(t^{-\gamma})f\|_\beta \leq ctt^{-1}\|f\|_\alpha \text{ (by (10.5))}$$

$$t\|S(t^{-\gamma})f\|_\beta \leq ct\|f\|_\beta \text{ (by (10.2))}$$

   (ii) Recall that by the fundamental lemma $t\frac{d}{dt}(D_K(t)) = D_J(t)$, and use (i).
   We now identify the Moser-Hormander spaces E_*^a as interpolation spaces.

Proposition 95 $E_*^a = (E^0, E^{a+1})_{\frac{a}{a+1},\infty;J}$.

Proof. In view of Lemma 94 (ii), if $f \in E_*^a$ then, with $D_J(t)f = t^{-\gamma}S'(t^{-\gamma})f$, $\gamma = 1/(a+1)$,

$$\|D_J(t)f\|_0 \leq cM \ t^{-\tau}t^{\gamma(a+1)},$$

$$\|D_J(t)f\|_{a+1} \leq cM \ t^{-\gamma}$$

Therefore,

$$t^{-a/(a+1)}J(t, D_J(t)f; E^0, E^{a+1}) \leq cMt^{-a/(a+1)}\max\{t^{a/(a+1)}, tt^{-1/(a+1)}\}$$

$$\leq cM$$

and taking the supremum over all $t \leq 1$, we obtain

$$\|f\|_{(E^0, E^{a+1})_{\frac{a}{a+1}, \infty; J}} \leq cM$$

Conversely, the argument also shows that $\|f\|_{(E^0, E^{a+1})_{\frac{a}{a+1}, \infty; J}} \leq M$ implies $\|f\|_{E_*^a} \leq cM$.

We say that a Banach scale $\{E^a\}_{a\geq 0}$ is *compact* if for every pair of indices $\alpha < \beta$ the injection $E^\beta \subset E^\alpha$ is compact. It is readily seen using Lemma 94 and the known theorems on interpolation of compactness that (cf. [49]): If there exists a pair of indices $\alpha_1 < \beta_1$ such that the embedding $E^{\beta_1} \subset E^{\alpha_1}$ is compact then the same is true for all the embeddings $E^\beta \subset E^\alpha$, $\alpha < \beta$.

Example 96 *(cf. [49]) Let $K \subset R^n$ be a compact set, let $\chi \in C_0^\infty(R^n)$ be such that $\chi \equiv 1$ in a neighborhood of K, and $\psi \in C_0^\infty(R^n)$ which is 1 in a neighborhood of the origin, and let $\hat{\varphi} = \psi$, $\varphi_\theta(x) = \theta^n\varphi(\theta x)$, $\theta \geq 1$. Set $S_\theta u = \chi(\varphi_\theta * u)$, then the scale of Holder spaces $\{H^a\}_{a\geq 0}$ is a Banach scale with smoothing given by the $\{S_\theta\}$.*

10.3 Abstract Nash-Moser Theorem

Let $\{E^a\}_{a\geq 0}$, $\{F^b\}_{b\geq 0}$ be Banach scales with their respective smoothing operators S_E, S_F. The following results are established in [49]. Let $\Phi : E^\infty \cap V \to F^\infty$, where V is a neighborhood of 0 in E_*^μ. Assume that:

i) Φ has a differential $\Phi'(u)$ for $u \in E^\infty \cap V$, and $\Phi'(u)$ has a right inverse $\psi(u)$ such that for certain α_1, α_2, with $0 \le \alpha_1 < \alpha < \alpha_2$ we have:

$$\|\psi(u)g\|_a \le c \sum (1 + \|u\|_{A_j(a)}) \|g\|_{B_j(a)}$$

where $\alpha_1 \le a \le \alpha_2$, $u \in E^\infty \cap V$, $g \in F^\infty$, the sum is finite, and $A_j(a)$, $B_j(a)$ are increasing linear functions.

ii) The map $(u, g) \to \psi(u)g$ is continuous from $E^\infty \cap V \to E^{\alpha_2}$.

iii) $\max\{\alpha_1, \mu\} < \alpha < \alpha_2$

iv) $B_j(a) \le \beta - \alpha + a$, for $\alpha_1 \le a \le \alpha_2$

v) $A_j(a) + B_j(a) < \alpha + \beta$, for $\alpha_1 \le a \le \alpha_2$

vi) The scale $\{E^a\}$ is compact.

Then $\psi(0)g \in E_*^\alpha$, if $g \in F_*^\beta$. Moreover, suppose that $\varepsilon > 0$ is small enough so that $A_j(a) + B_j(a) + \varepsilon < \alpha + \beta$, then there exists a neighborhood W_ε of 0 in $E_*^{\alpha-\varepsilon}$, such that if $u \in W_\varepsilon$, $g \in F_*^\beta$, then the operator

$$T_\psi(u)g = \int_1^\infty \psi(S_E(t)u)S_F'(t)g \, dt$$

is well defined, and the equation

$$u - \psi(0)g = (T_\psi(u) - T_\psi(0))g$$

has a solution $u \in W_\varepsilon$ for all $g \in F_*^\beta$ with sufficiently small norm. The crucial issues here are: i) to use the assumptions in order to establish the mapping properties of T_ψ and ii) to use the compactness together with the Leray-Schauder fixed point theorem to actually solve the equation.

Suppose further that

vii) $\Phi(0) = 0$

viii) $\|(\Phi'(u) - \Phi'(v))w\|_\beta \le c \sum (1 + \|u\|_{m_j} + \|v\|_{m_j'}) \|u-v\|_{m_j''} \|w\|_{m_j'''}$, for $u, v \in E^{\alpha_2} \cap V$, $w \in E^\infty$, where m_j', m_j'', $m_j''' \in [\alpha_1, \alpha_2]$

ix) $\alpha > \max\{m_j'' + m_j'''\}/2$, $\alpha > \max\{m_j' + m_j'' + m_j'''\}/3$

Then for every $f \in F_*^\beta$ with sufficiently small norm one can find $u \in E_*^\alpha$ with small norm such that $u\,(t) = \psi(0)S_F(1)g + \int_1^t \psi(S_E(s)u)S_F'(s)g \, ds \to$ to u (in E_*^α) and $\Phi(u\,(t)) \to f$ (in F_*^β) as $t \to \infty$, where the convergence is strong in E^a (resp F^b) when $a < \alpha$ (resp $b < \beta$).

At this stage it is routine to generalize the results of [49] for more general interpolation scales of functors with characteristic functions given by quasi concave functions more general than powers. Since we do not have any application for such results at this time we shall leave these developments for another occasion.

Bibliography

[1] H. Amann, *Parabolic evolution equations in interpolation and extrapolation spaces*, J. Funct. Anal. **78** (1988), 233-270.

[2] H. Amann, M. Milman, in preparation.

[3] D. Bakry, P. A. Meyer, *Sur les inegalites de Sobolev logarithmiques I & II*, Seminaire de Probabilités **16**, Lecture Notes in Math. **920** (1982), 146-160.

[4] J. M. Ball, *Convexity conditions and existence theorems in nonlinear elasticity*, Arch. Ration. Mech. Anal. **63** (1977), 337-403.

[5] T. Beale, T. Kato, A. Majda, *Remarks on the breakdown of smooth solutions for the 3 − d Euler equations*, Comm. Math. Phys. **94** (1984), 61-66.

[6] C. Bennett, K. Rudnick, *On Lorentz-Zygmund spaces*, Diss. Math. **175** (1980), 1-72.

[7] C. Bennett, R. Sharpley, *Interpolation of Operators*, Academic Press, 1988.

[8] J. Bergh, J. Löfström, *Interpolation spaces: An Introduction*, Springer-Verlag, Berlin-Heidelberg-New York (1976).

[9] H. Brezis, N. Fusco, C. Sbordone, *Integrability for the Jacobian of orientation preserving maps*, J. Functional Analysis, to appear.

[10] H. Brezis, T. Gallouet, *Nonlinear Schrödinger evolution equations*, J. Nonlin. Anal. **4** (1980), 677-681.

[11] H. Brezis, S. Wainger, *A note on limiting cases of Sobolev embeddings and convolution inequalities,* Comm. Part. Diff. Equations **5** (1980), 773-789.

[12] Ju. A. Brudnyĭ, N. Ja. Krugljak, *Interpolation functors and interpolation spaces* vol 1, North Holland, 1991.

[13] P. L. Butzer, H. Behrens, *Semigroups of operators and approximation,* Springer-Verlag, 1967.

[14] A. P. Calderón, *Intermediate spaces and interpolation, the complex method,* Studia Math. **24** (1964), 113-190.

[15] C. Calderón, M. Milman, *Interpolation of Sobolev spaces: the real method,* Indiana Math. J. **32** (1983), 801-808.

[16] P. Cannarsa, V. Vespri, *Generation of analytic semigroups in the L^p topology by elliptic operators in R^n,* Israel J. Math. **61** (1988), 235-255.

[17] L. Carleson, *On the convergence and growth of partial sums of Fourier series,* Acta Math. **116** (1966), 135-157.

[18] F. Cobos, M. Milman, *On a limit class of approximation spaces,* Numer. Funct. Anal. and Optimiz. **11** (1990), 11-31.

[19] R. Coifman, P. Lions, Y. Meyer, S. Semmes, *Compacité par compensation et éspaces de Hardy,* Comptes Rend. Acad. Sci. Paris **309** (1989), 945-949.

[20] R. Coifman, Y. Meyer, *Au delà des opérateurs pseudodifférentiels,* Astérisque **57** (1978).

[21] R. Coifman, A. McIntosh, Y. Meyer, *L'intégrale de Cauchy définit un opérateur borné sur L^2 pour les courbes lipschitziennes,* Ann. of Math. **116** (1982), 361-387.

[22] P. Constantin, *Collective L^∞ estimates for families of functions with orthormal derivatives,* Indiana Univ. Math. J. **36** (1987), 603-616.

[23] M. Cowling, I. Doust, A. McIntosch, A. Yagi, *Banach space operators with bounded H^∞ Functional Calculus*, preprint, 1993.

[24] M. Cwikel, B. Jawerth, M. Milman, *On the fundamental lemma of interpolation theory*, J. Approx. Theory **60** (1990), 70-82.

[25] M. Cwikel, B. Jawerth, M. Milman, R. Rochberg, *Differential estimates and commutators in interpolation theory*, Analysis at Urbana II, London Math. Soc. **138**, Cambridge Univ. Press 1989, 170-220.

[26] M. Cwikel, M. Milman, R. Rochberg (editors), *Interpolation Spaces and Related Topics*, Israel Math. Conference Proc. **5**, American Mathematical Society, 1992.

[27] M. Cwikel, M. Milman, J. Peetre, *Complex extrapolation*, preprint.

[28] M. Cwikel, *Weak type estimates for singular values and the number of bound states of Schrodinger operators*, Ann. Math. **106** (1977), 93-102.

[29] M. Cwikel, *K-divisibility of the K functional and Calderón couples*, Ark. Mat. **22** (1984), 39-62.

[30] M. Cwikel, P. Nilsson, *Interpolation of weighted Banach Lattices*, Memoirs Amer. Math. Soc., to appear.

[31] G. Da Prato, P. Grisvard, *Equation d'evolution abstraites non lineaires de type parabolique*, Ann. Mat. Pura Appl. **120** (1979), 329-396.

[32] G. Da Prato, P. Grisvard, *Sommes d'operateurs non lineaires et equations differentielles operationales*, J. Math. Pures Appl. **54** (1975), 305-387.

[33] G. Da Prato, P. Grisvard, *Maximal regularity for evolution equations by interpolation and extrapolation*, J. Functional Analysis **58** (1984), 107-124.

[34] R. DeVore, K. Scherer, *Interpolation of linear operators on Sobolev spaces*, Ann. Math. **109** (1979), 583-609.

[35] R. DeVore, B. Jawerth, V. Popov, *Compression of Wavelet decompositions*, to appear.

[36] G. Di Blasio, *Linear abstract parabolic equations in L^p spaces*, Ann. Mat. Pura Appl. **IV** (1984), 55-104.

[37] G. Di Blasio, *Characterization of interpolation spaces and regularity properties for holomorphic semigroups*, Semigroup Forum **38** (1989), 179-187.

[38] D. Donoho, *Unconditional bases are optimal for data compression and Statistical estimation*, Stanford, preprint 1993.

[39] G. Dore, H^∞ *Functional Calculus in Real Interpolation Spaces*, preprint, Univ. Bologna, 1993.

[40] D. E. Edmunds, P. Gurka, B. Opic, *Double exponential integrability of convolution operators in generalized Lorentz-Zygmund spaces*, Akad. Sc. Czech Republic, preprint, 1993.

[41] G. Feissner, *Hypercontractive semigroups and Sobolev's inequality*, Trans. Amer. Soc. **210** (1975), 51-62.

[42] N. Fusco, P. L. Lions, C. Sbordone, *Some remarks on Sobolev imbeddings in borderline cases*, preprint, Univ. Naples, 1993.

[43] M. Gomez, M. Milman, *Extrapolation spaces and a.e. convergence of singular integrals*, J. London Math. Soc. **34** (1986), 305-316.

[44] L. Greco, T. Iwaniec, *New Inequalities for the Jacobian*, Analyse Non-Linéaire, to appear.

[45] L. Greco, T. Iwaniec, M. Milman, in preparation.

[46] L. Gross, *Logarithmic Sobolev inequalities*, Amer. J. Math. **97** (1975), 1061-1083.

[47] R. S. Hamilton, *The inverse function theorem of Nash and Moser*, Bull. Amer. Math. Soc. **7** (1982), 65-222.

[48] R. Hemasinha, *Reiteration and duality of extrapolation spaces*, preprint.

[49] T. Holmstedt, *The equivalence of two methods of interpolation*, Math. Scand. **18** (1966), 45-62.

[50] L. Hörmander, *The Nash-Moser theorem and paradifferential operators*, Analysis, Etcetera, Academic Press, 1990.

[51] C. Houdre, manuscript in preparation.

[52] R. Hunt, *On the convergence of Fourier series*, Proc. Conf. Southern Illinois University (1967), 235-255.

[53] T. Iwaniec, A. Lutoborski, *Integral estimates for null lagrangians*, to appear.

[54] T. Iwaniec, C. Sbordone, *On the integrability of the Jacobian under minimal hypothesis*, Arch. Rational Mech. Anal. **119** (1992), 129-143.

[55] T. Iwaniec, C. Sbordone, *Weak minima of variational integrals*, to appear.

[56] B. Jawerth, M. Milman, *Interpolation of weak type spaces*, Math. Z. **201** (1989), 509-519.

[57] B. Jawerth, M. Milman, *Extrapolation theory with applications*, Memoirs Amer. Math. Soc. **440** (1991).

[58] B. Jawerth, M. Milman, *New results and applications of extrapolation theory*, Interpolation Spaces and Related Topics, (M. Cwikel, M. Milman, and R. Rochberg, editors), Israel Math. Confer. Proc. **5**, 1992.

[59] B. Jawerth, M. Milman, *Wavelets and best approximation in Besov spaces*, Interpolation Spaces and Related Topics, (M. Cwikel, M. Milman, and R. Rochberg, editors), Israel Math. Confer. Proc. **5**, 1992.

[60] B. Jawerth, R. Rochberg, G. Weiss, *Commutators and other second order estimates in interpolation theory*, Ark. Mat. **24** (1986), 191-219.

[61] N. Kalton, *Nonlinear commutators in interpolation theory*, Memoirs Amer. Math. Soc. **373** (1988).

[62] N. Kalton, *Differentials of complex interpolation processes for Köthe function spaces*, Transactions Amer. Math. Soc. 1992.

[63] T. Kato, G. Ponce, *Well posedness of the Euler and Navier Stokes equations in Lebesgue spaces $L_s^p(R^2)$*, Revista Matematica Iberoamericana **2** (1986), 73-88.

[64] R. Kerman, *An integral extrapolation theorem with applications*, Studia Math. **76** (1983), 183-195.

[65] H. König, *Eigenvalue distribution of compact operators*, Operator Theory Advances and Applications vol 16, Birkhäuser, 1986.

[66] A. Krasnoselski, P. E. Sobolevski, *Fractional powers of operators acting in Banach spaces*, Dokl. Nauk SSSR **129** (1959), 499-502.

[67] S. G. Krein, Ju Petunin, E. M. Semenov, *Interpolation of Linear Operators*, Transl. Math. Monogr. **54**, American Mathematical Society, 1982.

[68] N. Krugljak, *Embedding Theorems,Interpolation of operators and the Nash-Moser implicit function theorem*, Dokl. Akad. Nauk SSSR **226** (1976), 771-773.

[69] C. Li, A. McIntosh, K. Zhang, *Higher Integrability and Reverse Hölder Inequalities*, preprint, Macquarie University, 1993.

[70] E. Lieb, *An L^p bound for the Riesz and Bessel potentials of orthonormal functions*, J. Functional Anal. **51** (1983), 159-165.

[71] G. G. Lorentz, T. Shimogaki, *Interpolation theorems for operators in function spaces*, J. Funct. Anal. **2** (1968), 31-51.

[72] M. Milman, *Extrapolation spaces and a.e. convergence of Fourier series*, Journal of Approximation Theory, to appear.

[73] M. Milman, *The computation of the K functional for couples of rearrangement invariant spaces*, Result. Math. **5** (1982), 174-176.

[74] M. Milman, *Integrability of the Jacobians of orientation preserving maps: interpolation methods*, Comptes Rend. Acad. Sc. Paris **317** (1993), 539-543.

[75] M. Milman, *Inequalities for Jacobians: Interpolation techniques*, Rev. Mat. Colomb. **27** (1993), 67-81.

[76] M. Milman, *A commutator theorem with applications*, Coll. Math., to appear.

[77] M. Milman, *Higher order commutators in the real method of interpolation*, submitted.

[78] M. Milman, *Extrapolation and Logarithmic Sobolev Inequalities*, to appear.

[79] M. Milman, *A general form of DeLa Valle Poussin criteria*, in preparation.

[80] M. Milman, T. Schonbek, *A note on second order estimates in interpolation theory and applications*, Proc. Amer. Math. Soc. **110** (1990), 961-969.

[81] J. Moser, *A rapidly convergent iteration method and nonlinear partial differential equations I*, Ann. Scuola Norm. Sup. Pisa **20** (1966), 265-315.

[82] S. Müller, *Higher integrability of determinants and weak convergence in L^1*, J. Reine Angew. Math. **412** (1990), 20-34.

[83] S. Müller, *Hardy space methods for nonlinear partial differential equations*, preprint, Bonn, 1993.

[84] R. O'Neil, Les fonctions conjugées et les intégrales frac-
     tionarires de la classe $L(Log)^s$, Compt. Rend. Acad. Sc. Paris
     263 (1966), 463-466.

[85] V. I. Ovchinnikov, *The method of Orbits in Interpolation The-
     ory*, Math. Reports **1** (1984), 349-515.

[86] J. Peetre, *Zur interpolation von operatorenraumen*, Arch.
     Math. (Basel) **21** (1970), 601-608.

[87] E. I. Pustylnik, *On functions of a positive operator*, Math.
     USSR-Sb. **47** (1984), 27-42.

[88] R. Rochberg, *Higher order estimates in Complex Interpolation
     Theory*, to appear.

[89] R. Rochberg, G. Weiss, *Derivatives of analytic families of Ba-
     nach spaces*, Ann. Math. **118** (1983), 315-347.

[90] Y. Sagher, *Real interpolation with weights*, Indiana Math. J.
     30 (1981), 113-121.

[91] E. Sawyer, *Boundedness of classical operators on classical
     Lorentz spaces*, Studia Math. **96** (1990), 145-158.

[92] B. Simon, *Trace ideals and their applications*, London Math
     Soc. Lect. Notes **35**, Cambridge Univ. Press, 1979.

[93] J. Simon, *Sobolev, Besov and Nikolskii Fractional Spaces:
     Imbeddings and comparisons for vector valued spaces on an
     interval*, Ann. di Matem. Pura ed Appl. **97** (1990), 117-148.

[94] P. Sjölin, *An inequality of Paley and convergence a.e. of Walsh
     Fourier series*, Ark. Mat. **7** (1969), 531- 570.

[95] P. Sjölin, *Two theorems on Fourier integrals and Fourier series*,
     Approximation and Function Spaces, Banach Center Publ. **22**
     (1989), 413-426.

[96] P. E. Sobolevskii, A. I. Yasakov, *The equations of parabolic type
     with nonpower nonlinearities*, Differential Equations **7** (1971).

[97] F. Soria, *Integrability properties of the maximal operator of partial sums of Fourier series*, Rendiconti del Circ. Mat. Palermo **38** (1989), 371-376.

[98] E. M. Stein, *Singular integrals and differentiability properties of functions*, Princeton Univ. Press, 1970.

[99] M. Taylor, *Pseudodifferential operators and nonlinear PDE*, Progress in Math. **100**, Birkhauser, 1991.

[100] R. Teman, *Infinite dimensional dynamical systems in mechanics and physics*, Applied Mathematical Sc. **68**, Springer, 1988.

[101] H. Triebel, *Interpolation theory, Function spaces, Differential Operators*, North Holland, Amsterdam, 1978.

[102] N. Varopoulos, *Hardy-Littlewood theory for semigroups*, J. Funct. Anal. **63** (1985), 240-260.

[103] M. Zafran, *On the symbolic calculus in homogeneous Banach algebras*, Israel J. Math. **32** (1979), 183-192.

[104] A. Zygmund, *Trigonometric Series*, Cambridge University Press, (1959)

Index

C
Calderon-Zygmund
    Operator, 40
conventions
    notation, 34

D
decomposition
    almost optimal, 98

E
estimates
    weighted norm, 31
exact
    exponent, 11
extrapolation
    exact, 8

F
F- method
    interpolation, 71
family
    strongly compatible, 8
function
    characteristic, 11
    presentable, 60
    quasi-concave, 11
    tempered, 18
functional
    J, 11
    K, 11

functor
    interpolation, 7
fundamental
    lemma, 14

G
Gagliardo
    completion, 13
    diagram, 13

H
Hilbert Transform, 29

I
ideals
    operator, 67
inequalities
    division of, 28
    K/J, 25
    rearrangement, 29
integrability
    double exponential, 91
interpolation functor
    characteristic function, 11
    complete, 25
    exact, 11
    of exponent, 11
    exact, 8
    space, 7

J
Jacobian, 47

K
K/J inequality, 20

M
Macaev ideals
    ideals of operators, 67
maps
    orientation preserving, 46
    Sobolev, 47
measure
    representing, 17
method
    extrapolation, 8
minorant
    largest logarithmically convex, 17

N
notation
    conventions, 34

O
Operator
    Calderon-Zygmund, 29, 40
orientation preserving
    map, 47

P
pair
    compatible, 7
    mutually closed, 13
    quasilinearizable, 142
    regular, 25

R
reiteration
    constancy property, 76

S
scale
    ordered, 9
    with smoothing, 143
Schatten classes
    ideal of operator, 67
semigroup
    Hermite-Ornstein-Uhlenbeck, 90
SFL, 14
Sobolev
    Logarithmic Inequalities, 89
    Spaces, 40
space
    exact, 7
    extrapolation, 8
    Hormander, 144
    intermediate, 7
    Lorentz, 23
    rearrangement invariant, 29
    Sobolev, 47
    Zafran, 37

W
weight, 99

Symbols

$\bar{A}$	2.1	$\Delta(.)$	2.1
$\bar{A}_{\theta,q;J}$	2.1	$\sum(.)$	2.1
$\bar{A}_{\theta,q;K}$	2.1	$\sum_p(.)$	2.1
$\bar{A}_{\theta,q}$	2.7	E_*^a	10.2
$\bar{A}_{\rho,1;J}$	2.1	$ExpL^\alpha$	2.2
$\bar{A}_{\rho,\infty;K}$	2.1	$\mathcal{F}(\bar{A})^0$	2.1
$\bar{A}_{\psi;K}$	7.6	$J(t,a;\bar{A})$	2.1
$\bar{A}_{(\alpha);J}$	2.2	$K(t,a;\bar{A})$	2.1
$\bar{A}_{(\alpha);K}$	2.2	$K_L(t,a;\bar{A})$	10.1
$\bar{A}_{(\alpha)}$	2.2	Λ_φ	2.2
$\bar{A}_f$	3.1	$L(LogL)^\alpha$	2.2
$\bar{A}_{0,q;K}$	6.1	$L^p(LogL)^\alpha$	6.4
D_A	9.1	$\Gamma(a)$	2.1
$D_J(t)$	7.1	W_p^k	4.2.2
$D_K(t)$	7.1	$W_p^k(\Omega, R^n)$	4.2.1
Ω, Ω_n	7.1	S_p, S_w, S_M	5.2

Printing: Weihert-Druck GmbH, Darmstadt
Binding: Theo Gansert Buchbinderei GmbH, Weinheim

Vol. 1488: A. Carboni, M. C. Pedicchio, G. Rosolini (Eds.), Category Theory. Proceedings, 1990. VII, 494 pages. 1991.

Vol. 1489: A. Mielke, Hamiltonian and Lagrangian Flows on Center Manifolds. X, 140 pages. 1991.

Vol. 1490: K. Metsch, Linear Spaces with Few Lines. XIII, 196 pages. 1991.

Vol. 1491: E. Lluis-Puebla, J.-L. Loday, H. Gillet, C. Soulé, V. Snaith, Higher Algebraic K-Theory: an overview. IX, 164 pages. 1992.

Vol. 1492: K. R. Wicks, Fractals and Hyperspaces. VIII, 168 pages. 1991.

Vol. 1493: E. Benoît (Ed.), Dynamic Bifurcations. Proceedings, Luminy 1990. VII, 219 pages. 1991.

Vol. 1494: M.-T. Cheng, X.-W. Zhou, D.-G. Deng (Eds.), Harmonic Analysis. Proceedings, 1988. IX, 226 pages. 1991.

Vol. 1495: J. M. Bony, G. Grubb, L. Hörmander, H. Komatsu, J. Sjöstrand, Microlocal Analysis and Applications. Montecatini Terme, 1989. Editors: L. Cattabriga, L. Rodino. VII, 349 pages. 1991.

Vol. 1496: C. Foias, B. Francis, J. W. Helton, H. Kwakernaak, J. B. Pearson, H∞-Control Theory. Como, 1990. Editors: E. Mosca, L. Pandolfi. VII, 336 pages. 1991.

Vol. 1497: G. T. Herman, A. K. Louis, F. Natterer (Eds.), Mathematical Methods in Tomography. Proceedings 1990. X, 268 pages. 1991.

Vol. 1498: R. Lang, Spectral Theory of Random Schrödinger Operators. X, 125 pages. 1991.

Vol. 1499: K. Taira, Boundary Value Problems and Markov Processes. IX, 132 pages. 1991.

Vol. 1500: J.-P. Serre, Lie Algebras and Lie Groups. VII, 168 pages. 1992.

Vol. 1501: A. De Masi, E. Presutti, Mathematical Methods for Hydrodynamic Limits. IX, 196 pages. 1991.

Vol. 1502: C. Simpson, Asymptotic Behavior of Monodromy. V, 139 pages. 1991.

Vol. 1503: S. Shokranian, The Selberg-Arthur Trace Formula (Lectures by J. Arthur). VII, 97 pages. 1991.

Vol. 1504: J. Cheeger, M. Gromov, C. Okonek, P. Pansu, Geometric Topology: Recent Developments. Editors: P. de Bartolomeis, F. Tricerri. VII, 197 pages. 1991.

Vol. 1505: K. Kajitani, T. Nishitani, The Hyperbolic Cauchy Problem. VII, 168 pages. 1991.

Vol. 1506: A. Buium, Differential Algebraic Groups of Finite Dimension. XV, 145 pages. 1992.

Vol. 1507: K. Hulek, T. Peternell, M. Schneider, F.-O. Schreyer (Eds.), Complex Algebraic Varieties. Proceedings, 1990. VII, 179 pages. 1992.

Vol. 1508: M. Vuorinen (Ed.), Quasiconformal Space Mappings. A Collection of Surveys 1960-1990. IX, 148 pages. 1992.

Vol. 1509: J. Aguadé, M. Castellet, F. R. Cohen (Eds.), Algebraic Topology - Homotopy and Group Cohomology. Proceedings, 1990. X, 330 pages. 1992.

Vol. 1510: P. P. Kulish (Ed.), Quantum Groups. Proceedings, 1990. XII, 398 pages. 1992.

Vol. 1511: B. S. Yadav, D. Singh (Eds.), Functional Analysis and Operator Theory. Proceedings, 1990. VIII, 223 pages. 1992.

Vol. 1512: L. M. Adleman, M.-D. A. Huang, Primality Testing and Abelian Varieties Over Finite Fields. VII, 142 pages. 1992.

Vol. 1513: L. S. Block, W. A. Coppel, Dynamics in One Dimension. VIII, 249 pages. 1992.

Vol. 1514: U. Krengel, K. Richter, V. Warstat (Eds.), Ergodic Theory and Related Topics III, Proceedings, 1990. VIII, 236 pages. 1992.

Vol. 1515: E. Ballico, F. Catanese, C. Ciliberto (Eds.), Classification of Irregular Varieties. Proceedings, 1990. VII, 149 pages. 1992.

Vol. 1516: R. A. Lorentz, Multivariate Birkhoff Interpolation. IX, 192 pages. 1992.

Vol. 1517: K. Keimel, W. Roth, Ordered Cones and Approximation. VI, 134 pages. 1992.

Vol. 1518: H. Stichtenoth, M. A. Tsfasman (Eds.), Coding Theory and Algebraic Geometry. Proceedings, 1991. VIII, 223 pages. 1992.

Vol. 1519: M. W. Short, The Primitive Soluble Permutation Groups of Degree less than 256. IX, 145 pages. 1992.

Vol. 1520: Yu. G. Borisovich, Yu. E. Gliklikh (Eds.), Global Analysis – Studies and Applications V. VII, 284 pages. 1992.

Vol. 1521: S. Busenberg, B. Forte, H. K. Kuiken, Mathematical Modelling of Industrial Process. Bari, 1990. Editors: V. Capasso, A. Fasano. VII, 162 pages. 1992.

Vol. 1522: J.-M. Delort, F. B. I. Transformation. VII, 101 pages. 1992.

Vol. 1523: W. Xue, Rings with Morita Duality. X, 168 pages. 1992.

Vol. 1524: M. Coste, L. Mahé, M.-F. Roy (Eds.), Real Algebraic Geometry. Proceedings, 1991. VIII, 418 pages. 1992.

Vol. 1525: C. Casacuberta, M. Castellet (Eds.), Mathematical Research Today and Tomorrow. VII, 112 pages. 1992.

Vol. 1526: J. Azéma, P. A. Meyer, M. Yor (Eds.), Séminaire de Probabilités XXVI. X, 633 pages. 1992.

Vol. 1527: M. I. Freidlin, J.-F. Le Gall, Ecole d'Eté de Probabilités de Saint-Flour XX – 1990. Editor: P. L. Hennequin. VIII, 244 pages. 1992.

Vol. 1528: G. Isac, Complementarity Problems. VI, 297 pages. 1992.

Vol. 1529: J. van Neerven, The Adjoint of a Semigroup of Linear Operators. X, 195 pages. 1992.

Vol. 1530: J. G. Heywood, K. Masuda, R. Rautmann, S. A. Solonnikov (Eds.), The Navier-Stokes Equations II – Theory and Numerical Methods. IX, 322 pages. 1992.

Vol. 1531: M. Stoer, Design of Survivable Networks. IV, 206 pages. 1992.

Vol. 1532: J. F. Colombeau, Multiplication of Distributions. X, 184 pages. 1992.

Vol. 1533: P. Jipsen, H. Rose, Varieties of Lattices. X, 162 pages. 1992.

Vol. 1534: C. Greither, Cyclic Galois Extensions of Commutative Rings. X, 145 pages. 1992.

Vol. 1535: A. B. Evans, Orthomorphism Graphs of Groups. VIII, 114 pages. 1992.

Vol. 1536: M. K. Kwong, A. Zettl, Norm Inequalities for Derivatives and Differences. VII, 150 pages. 1992.

Vol. 1537: P. Fitzpatrick, M. Martelli, J. Mawhin, R. Nussbaum, Topological Methods for Ordinary Differential Equations. Montecatini Terme, 1991. Editors: M. Furi, P. Zecca. VII, 218 pages. 1993.

Vol. 1538: P.-A. Meyer, Quantum Probability for Probabilists. X, 287 pages. 1993.

Vol. 1539: M. Coornaert, A. Papadopoulos, Symbolic Dynamics and Hyperbolic Groups. VIII, 138 pages. 1993.

Vol. 1540: H. Komatsu (Ed.), Functional Analysis and Related Topics, 1991. Proceedings. XXI, 413 pages. 1993.

Vol. 1541: D. A. Dawson, B. Maisonneuve, J. Spencer, Ecole d´ Eté de Probabilités de Saint-Flour XXI - 1991. Editor: P. L. Hennequin. VIII, 356 pages. 1993.

Vol. 1542: J.Fröhlich, Th.Kerler, Quantum Groups, Quantum Categories and Quantum Field Theory. VII, 431 pages. 1993.

Vol. 1543: A. L. Dontchev, T. Zolezzi, Well-Posed Optimization Problems. XII, 421 pages. 1993.

Vol. 1544: M.Schürmann, White Noise on Bialgebras. VII, 146 pages. 1993.

Vol. 1545: J. Morgan, K. O'Grady, Differential Topology of Complex Surfaces. VIII, 224 pages. 1993.

Vol. 1546: V. V. Kalashnikov, V. M. Zolotarev (Eds.), Stability Problems for Stochastic Models. Proceedings, 1991. VIII, 229 pages. 1993.

Vol. 1547: P. Harmand, D. Werner, W. Werner, M-ideals in Banach Spaces and Banach Algebras. VIII, 387 pages. 1993.

Vol. 1548: T. Urabe, Dynkin Graphs and Quadrilateral Singularities. VI, 233 pages. 1993.

Vol. 1549: G. Vainikko, Multidimensional Weakly Singular Integral Equations. XI, 159 pages. 1993.

Vol. 1550: A. A. Gonchar, E. B. Saff (Eds.), Methods of Approximation Theory in Complex Analysis and Mathematical Physics IV, 222 pages. 1993.

Vol. 1551: L. Arkeryd, P. L. Lions, P.A. Markowich, S.R. S. Varadhan. Nonequilibrium Problems in Many-Particle Systems. Montecatini, 1992. Editors: C. Cercignani, M. Pulvirenti. VII, 158 pages 1993.

Vol. 1552: J. Hilgert, K.-H. Neeb, Lie Semigroups and their Applications. XII, 315 pages. 1993.

Vol. 1553: J.-L- Colliot-Thélène, J. Kato, P. Vojta. Arithmetic Algebraic Geometry. Trento, 1991. Editor: E. Ballico. VII, 223 pages. 1993.

Vol. 1554: A. K. Lenstra, H. W. Lenstra, Jr. (Eds.), The Development of the Number Field Sieve. VIII, 131 pages. 1993.

Vol. 1555: O. Liess, Conical Refraction and Higher Microlocalization. X, 389 pages. 1993.

Vol. 1556: S. B. Kuksin, Nearly Integrable Infinite-Dimensional Hamiltonian Systems. XXVII, 101 pages. 1993.

Vol. 1557: J. Azéma, P. A. Meyer, M. Yor (Eds.), Séminaire de Probabilités XXVII. VI, 327 pages. 1993.

Vol. 1558: T. J. Bridges, J. E. Furter, Singularity Theory and Equivariant Symplectic Maps. VI, 226 pages. 1993.

Vol. 1559: V. G. Sprindžuk, Classical Diophantine Equations. XII, 228 pages. 1993.

Vol. 1560: T. Bartsch, Topological Methods for Variational Problems with Symmetries. X, 152 pages. 1993.

Vol. 1561: I. S. Molchanov, Limit Theorems for Unions of Random Closed Sets. X, 157 pages. 1993.

Vol. 1562: G. Harder, Eisensteinkohomologie und die Konstruktion gemischter Motive. XX, 184 pages. 1993.

Vol. 1563: E. Fabes, M. Fukushima, L. Gross, C. Kenig, M. Röckner, D. W. Stroock, Dirichlet Forms. Varenna, 1992. Editors: G. Dell'Antonio, U. Mosco. VII, 245 pages. 1993.

Vol. 1564: J. Jorgenson, S. Lang, Basic Analysis of Regularized Series and Products. IX, 122 pages. 1993.

Vol. 1565: L. Boutet de Monvel, C. De Concini, C. Procesi, P. Schapira, M. Vergne. D-modules, Representation Theory, and Quantum Groups. Venezia, 1992. Editors: G. Zampieri, A. D'Agnolo. VII, 217 pages. 1993.

Vol. 1566: B. Edixhoven, J.-H. Evertse (Eds.), Diophantine Approximation and Abelian Varieties. XIII, 127 pages. 1993.

Vol. 1567: R. L. Dobrushin, S. Kusuoka, Statistical Mechanics and Fractals. VII, 98 pages. 1993.

Vol. 1568: F. Weisz, Martingale Hardy Spaces and their Application in Fourier Analysis. VIII, 217 pages. 1994.

Vol. 1569: V. Totik, Weighted Approximation with Varying Weight. VI, 117 pages. 1994.

Vol. 1570: R. deLaubenfels, Existence Families, Functional Calculi and Evolution Equations. XV, 234 pages. 1994.

Vol. 1571: S. Yu. Pilyugin, The Space of Dynamical Systems with the C^0-Topology. X, 188 pages. 1994.

Vol. 1572: L. Göttsche, Hilbert Schemes of Zero-Dimensional Subschemes of Smooth Varieties. IX, 196 pages. 1994.

Vol. 1573: V. P. Havin, N. K. Nikolski (Eds.), Linear and Complex Analysis – Problem Book 3 – Part I. XXII, 489 pages. 1994.

Vol. 1574: V. P. Havin, N. K. Nikolski (Eds.), Linear and Complex Analysis – Problem Book 3 – Part II. XXII, 507 pages. 1994.

Vol. 1575: M. Mitrea, Clifford Wavelets, Singular Integrals, and Hardy Spaces. XI, 116 pages. 1994.

Vol. 1576: K. Kitahara, Spaces of Approximating Functions with Haar-Like Conditions. X, 110 pages. 1994.

Vol. 1577: N. Obata, White Noise Calculus and Fock Space. X, 183 pages. 1994.

Vol. 1358: D. Mumford, The Red Book of Varieties and Schemes. 2nd Printing. VII, 310 pages. 1994.

Vol. 1578: J. Bernstein, V. Lunts, Equivariant Sheaves and Functors. V, 139 pages. 1994.

Vol. 1579: N. Kazamaki, Continuous Exponential Martingales and BMO. VII, 91 pages. 1994.

Vol. 1580: M. Milman, Extrapolation and Optimal Decompositions with Applications to Analysis. XI, 161 pages. 1994.

Vol. 1581: D. Bakry, R. D. Gill, S. A. Mochanov, Lectures on Probability Theory. Editor: P. Bernard. VIII, 420 pages. 1994.